Mapping the
North Carolina Coast

Mapping the North Carolina Coast

Sixteenth-Century Cartography and the Roanoke Voyages

William P. Cumming

America's Four Hundredth Anniversary Committee

Lindsay C. Warren, Jr.
Chairman

Marc Basnight
Luther Daniels
Mrs. Burwell Evans
Andy Griffith

Robert V. Owens, Jr.
William S. Powell
L. Richardson Preyer
S. Thomas Rhodes
Mrs. J. Emmett Winslow

David Stick
Mrs. Percy Tillett
Charles B. Wade, Jr.
Charles B. Winberry, Jr.

John D. Neville
Executive Director

Mrs. Marsden B. deRosset, Jr.
Assistant Director

Advisory Committee on Publications

William S. Powell
Chairman

Lindley S. Butler
Jerry C. Cashion
David Stick
Alan D. Watson

ISBN 0-8652-232-2

Contents

List of Maps

Foreword

America's Four Hundredth Anniversary Committee, formed in 1978 under the provisions of an act of the North Carolina General Assembly of 1973, was charged with recommending plans for the observance of the quadricentennial of the first English attempts to explore and settle North America. The committee has proposed to carry out a variety of programs to appeal to a broad range of people. Among these is a publications program that includes a series of booklets dealing with the history of the events and people of the 1580s.

Queen Elizabeth I of England enjoyed a reign that was for the most part peaceful. It was a period of prosperity, which saw the flourishing of a new interest in literature, religion, exploration, and business. English mariners began to venture farther from home, and in time talk began to be heard of hopes to establish naval bases and colonies in America. Men of the County of Devon in the southwest of England, seafarers for generations, played leading roles in this expansion. One of these, Walter Ralegh (as he most often wrote his name), became a favorite of the queen, and on him she bestowed a variety of honors and rewards. It was he to whom she granted a charter in 1584 authorizing the discovery and occupation of lands not already held by "any Christian Prince and . . . people." Ralegh promptly sent a reconnaissance expedition to what is now North Carolina, and this was followed in due time by a colony under the leadership of Ralph Lane. Headquarters were established on Roanoke Island. After remaining for nearly a year and exploring far afield, Lane and his men returned to England in 1586.

In the summer of 1587 Governor John White and a colony of 115 men, women, and children arrived and occupied the houses and the fort left by Lane. The brief annals of this colony are recorded in a journal kept by the governor; they tell of certain problems that arose early—but they also record the birth of the first English child in America. The journal further explains why Governor White consented to return to England for supplies. His departure was the last contact with the settlers who constituted the "Lost Colony," renowned in history, literature, and folklore.

Although a casual acquaintance with the facts of these English efforts might suggest that they were failures, such was far from the case. Ralegh's expenditures of time, effort, and resources (in which he was joined by many others, including Queen Elizabeth herself) had salutary effects for England and certainly for all of present-day America. From Ralegh's initial investment in the reconnaissance voyage, as well as from the colonies, came careful descriptions of the New World and samples of its products. The people of England, indeed of the Western world, learned about North America; because books were published based on what Ralegh's men discovered, they could soon read for themselves of the natives there and the promise of strange and wonderful new resources.

From these voyages and colonizing efforts came the conviction that an English nation could be established in America. In 1606, when another charter was about to be issued for further settlement, King James, who succeeded Queen Elizabeth at her death in 1603, called for advice from some of the men who had been associated with Ralegh. They assured the king that further efforts would surely succeed. With this the Virginia Company was chartered, and it established England's first permanent settlement in America at Jamestown.

Because of Sir Walter Ralegh's vision, England persisted. Because of England's persistence and its refusal to yield to Spain's claims to the region, the United States today enjoys an English heritage. The English common law is the basis of American law; American legislative bodies are modeled on the House of Commons with the rights and freedoms that it developed over a long period of time; America's mother tongue is English, and it is the most commonly spoken language in the world—pilots and navigators on international airlines and the controllers who direct them at airports all over the world use English. Americans also share England's literary tradition: Chaucer, Beowulf, King Arthur, and Shakespeare are America's too, and Americans can enjoy Dickens and Tennyson, as well as Agatha Christie and Dorothy Sayers. America's religious freedom is also in the English tradition, and several of this nation's Protestant denominations trace their earliest history to origins in England: the Episcopal church, certainly, but the Quakers, Baptists, Congregationalists, and Universalists as well.

America's Four Hundredth Anniversary Committee has planned many programs to direct national and even international attention to the significance of events that occurred from bases established by English men, women, and children, but notably Sir Walter Ralegh, in what is now North Carolina during the period 1584-1590. While some of the programs may be regarded as fleeting and soon forgotten, the publications are intended to serve as lasting reminders of America's indebtedness to England. Books, pamphlets, and folders covering a broad range of topics have been prepared by authors on both sides of the Atlantic. These, it is anticipated, will introduce a vast new audience to the facts of America's origins.

Lindsay C. Warren, Jr., *Chairman*
America's Four Hundredth Anniversary Committee

Acknowledgments

To many scholars and librarians in the United States and abroad I owe thanks for information given and for unfailing courtesy. Especially am I indebted to two old friends and collaborators on other subjects: To David Beers Quinn I am indebted, as is every modern writer who has dealt with early North American exploration and in particular the Roanoke voyages, for his tenacity and adroitness in discovering documentary sources, his judgment in interpreting them, and his felicity in commenting on them. Also to another friend, Helen Wallis, I am grateful for help in years spent in the British Museum and British Library and for her corrections and suggested changes in the present work, fresh from her own knowledge of relevant material in preparing the catalog for the "Raleigh & Roanoke" exhibit. Finally, I am indebted to my wife and research assistant, Elizabeth, and to my son, Robert, for reading with experienced editorial eyes and commenting on the earlier draft of the work. To these, for their suggestions and corrections, I, as well as the reader, am in debt; the errors are mine.

William P. Cumming

I. The Beginnings

On a midsummer's day some four hundred years ago, two small Elizabethan vessels approached a low-lying, islanded coast. The ships had left the nearby Gulf Stream, a navigational route well known to Spanish Cadiz-bound treasure fleets, as well as to English and French pirates and privateers who pursued them or ravaged the Caribbean ports. These two English vessels were on a different mission, however: that of finding a suitable location for a colony.

Arthur Barlowe, captain of the smaller ship and a gentleman who had served as a soldier with Sir Walter Ralegh in Ireland, left the only extant narration of what the crewmen of these vessels saw:

> The first voyage made to the coasts of America, with two barks, wherein were Captaines Master Philip Amadas and Master Arthur Barlowe, who discovered part of the countrey, now called Virginia, anno 1584. . . .
>
> The second of July, we found shole water, which smelt so sweetly, and was so strong a smell, as if we had bene in the midst of some delicate garden . . . by which we were assured that land could not be farre distant: and keeping good watch and but slacke saile, the fourth of the same month, we arrived upon the coast, which we supposed to be a continent, and firme lande, and wee sailed along the same, a hundred and twentie English miles, before we could finde any entrance, or river issuing into the Sea.[1]

Thus Captain Barlowe described the arrival on the Carolina Outer Banks of an exploratory voyage sent by Sir Walter Ralegh to search for a suitable location for his proposed colony in a wide territory granted him earlier that year by Queen Elizabeth I. By "the first voyage made to the coasts of America," Barlowe was referring only to the first reconnaissance made for Ralegh's colony. Many other European ships had sailed along the same coastlines. The crews of some had landed willingly, others unwillingly or disastrously, on the sandy shoals and shallow inlets that form one of the most dangerous and turbulent regions along the North Atlantic coast. The Barlowe crew's knowledge of some of these previous voyages and of the maps and reports that resulted from them explains the choice of this area for the colony and is crucial to the understanding of the motivations leading to it. The following pages explore the cartography known or available to the English colonists and that which they produced.

It was not, however, from the earliest maps that the Ralegh colonists gained useful geographic information. For more than twenty-five years after the discovery of the West Indies, maps continued to present confused and divergent ideas of what lay north of Cuba: a blank area, Asia, or vaguely delineated islands or landmasses. The earliest dated and surviving map of the world to show the discoveries in the New World is that by Juan de la Cosa, a mariner and cartographer who sailed with Christopher Columbus on his second crossing in 1493-1494 and made three later voyages.[2] This map, dated 1500, shows to the north of Cuba a broken, vaguely formed coast with a long east-west line for the Newfoundland-Cape Breton region. Below it is the legend "Sea discovered by the English," with numerous coastal place-names; without question it reports the 1497 voyage of John Cabot for King Henry VII of England. The mainland, trending southwestward in somewhat stylized cusps, has no nomenclature, and the indentations on the coast are unidentifiable. This does not mean that it is entirely imaginary, though Cosa may have thought he was portraying Asia. (Columbus had required his officers and crew to declare upon oath that they had reached Asia.)

Additional western voyages of exploration by Spaniards, authorized by King Ferdinand and Queen Isabella, were being made before 1500. After Ferdinand granted Columbus's monopoly, surreptitious or unauthorized expeditions were also undertaken, from which no official report could be expected; and storms may have driven ships off their intended courses. Enterprises originating from ports not under Spanish control also were making exploratory transatlantic crossings; one such example is John Cabot's voyage of 1498, from which he never returned to England. The lack of surviving reports and maps does not mean, however, that contemporaries did not hear from or about them.

Such a "free-lance" voyage or voyages must have been the source for the idea of a landmass north of Cuba that became a dominant feature on maps during the first two decades of the sixteenth century. Such a landmass is first seen on a chart made in Lisbon in 1502 for the duke of Ferrara by order of his Portuguese envoy, Alberto Cantino. The identification of this unnamed landmass, which extends north of the island of "Ysabella" (usually accepted as Cuba), is a baffling and still hotly disputed problem, and it illustrates well the difficulty of interpreting geographic delineations that are themselves interpretations made by European cartographers and geographers of what they had obtained from returning travelers. Some recent scholars claim that the landmass shown above Cuba is the Yucatan peninsula, misplaced by mistake; others argue vigorously that it is Cuba duplicated by error. A third interpretation is that it is Florida, correctly trending northwest by north on its east coast and then continuing northward for a thousand miles. The islands west of its southern point would be Key West and the Ten Thousand Islands. This last interpretation is both the most obvious and, on the basis of present evidence, the most reasonable.[3]

The most widely known maps and charts showing this feature are the printed maps of Martin Waldseemüller, a learned monk of St. Dié, Switzerland, which include a great wall map of the world (1507) in twelve sheets (preserved in only one copy, in Wolfegg Castle, Germany); and several smaller maps in various edi-

2

tions of Ptolemy and elsewhere. Many variations of this chart, in manuscript and printed, appeared for many years.

In the 1520s voyages of exploration along the present Carolina coast occurred. These voyages resulted in reports and maps that, in divergent ways, influenced European conceptions of the region and resulted in attempts at settlement. To the north, Newfoundland and the Banks were already attracting fishing fleets from Portuguese, Basque, French, and British ports; to the south, Ponce de Leon had discovered and coasted along both sides of the Florida peninsula. By 1520 voyages around the entire perimeter of the Caribbean and the Atlantic coast of the South American continent revealed a continuous land barrier to the South Sea (the Pacific) and to the riches of the Indies, China, and Japan.[4] Then, on September 6, 1522, aboard the *Vittoria* (which also carried a rich cargo of spices from the East), Juan Sebastian del Cano reached San Lucar on the southwest coast of Spain with the survivors of Ferdinand Magellan's fleet and reported the first circumnavigation of the world.

The news of Cano's achievement spread rapidly through western Europe. Pigafetta, Magellan's comrade and the historian of the voyage, had already visited the courts of Spain, Portugal, and France before his return to his home in Venice by the autumn of 1523, and two French accounts of the Magellan voyage had been published in 1523.[5] But the length of the circuitous route around South America and the tempestuous hazards of over 300 miles through the Straits of Tierra del Fuego raised thoughts and hopes of a more direct passage to the East.

3

II. The Voyage of Verrazzano

By 1523 France, which alone among the four great European powers bordering the Atlantic had shown no active concern for western exploration, began to exhibit both private and royal interest. In Lyons, at the head of shipping on the Rhone River, were important Italian banking and commercial concerns that had grown into a major branch of the powerful Guadagni banking firm of Florence, as well as a distribution point for silk and spices from China and the Levant. The trade of these concerns had been severely damaged by the Portuguese discovery of a water route to the East by way of the Cape of Good Hope and the Indian Ocean. If a shorter route by a westward passage to China and the East were found, the way would be open to competitive trade from French ports. A small group within this powerful and influential "Florentine Nation of Lyons" pledged financial support for a westward voyage of discovery, to be led by a Florentine pilot, Giovanni da Verrazzano.[6] Verrazzano's voyage of 1523-1524 resulted in a narrative, a map, and long-lasting influence on other maps. In addition, it was of major significance to the plans and actions of the Ralegh colonists.

Little is known about Verrazzano's life before 1523. He was apparently connected with the well-established Verrazzano family that had had close banking affiliations in Florence and Lyons, and he was well educated in the Florentine humanistic tradition. Contemporary references point to his years in Cairo or the Levant, to his wide maritime experience, and to his having sailed with Magellan from Portugal to Spain, though there is no explanation of why he did not, like his fellow countryman Pigafetta, continue on the voyage of circumnavigation. It is clear that Verrazzano was an excellent mathematician and was highly regarded by his associates in Lyons, as well as by the king of France, Francis I.

The king's assent was necessary before Verrazzano's proposed exploration was feasible, but the explorer could broach his plan under conditions at least temporarily favorable. The considerable reputation of Francis I had resulted from his achievements on the battlefield and in the boudoir, but his aggressive military activities with and against his neighbors and the maintenance of the most magnificent court in Europe had put him deeply in need of money. The capture of a French corsair of three galleons sent by the Spanish explorer Hernando Cortés with some of the fabulous treasure of Montezuma, Aztec emperor of Mexico; the reports of the fishing grounds off Newfoundland, to which Breton fishing crews had been making their way since 1504 or before; and the quest for a way to the wealth of the East must have stimulated the king's imagination.

Francis openly expressed his contempt for the division of the non-Christian world between Spain and Portugal in the Treaty of Tordesillas (1494). This treaty drew a line of demarcation at a meridian about 370 leagues west of the Cape Verde Islands (about 51 degrees west longitude); Spain owned all lands west of this line toward India not held by a Christian king on Christmas Day, 1492; Portugal, all lands eastward. Both countries claimed legal and papal authority for such hegemony and were therefore opposed to any intrusion by another European power for the purpose of entering or establishing relations—especially commercial—in their respective spheres of influence.

France was already at war with Spain; the Portuguese envoy to the French king protested vigorously but unavailingly any approval of Verrazzano's voyage. Francis I, openly disregarding the claims of both countries, authorized the voyage and assigned to it the *Dauphine,* an armed 100-ton caravel of the royal navy, and paid from the royal treasury the salary of its captain, Antoine de Conflans. The *Dauphine* had a crew of fifty.

Upon his return Verrazzano wrote the king a literate and detailed report. The original report is apparently lost, but its contents are preserved in several varying versions. Giovanni Battista Ramusio published a somewhat shortened version of Verrazzano's report in Italian in the third volume of his *Navigationi et viaggi* (Venice, 1556). This publication was translated into English by Richard Hakluyt and included in his *Diuers voyages* (London, 1582); thus, it was made available to those planning or making the Ralegh voyages. The fullest and best extant copy of the royal report, with valuable marginal notes generally accepted as being in Verrazzano's own hand (and found in no other version) is now in the J. P. Morgan Library in New York City. The passages from the letter quoted here are from the Cèllere Codex, as the Morgan manuscript is called (from the name of the previous owner). Two maps that record firsthand information probably furnished by Verrazzano, or from the ship's journal and sketches, are a map of the world by Vesconte de Maggiolo dated 1527 and another world chart by Giovanni's brother, Gerolamo (Hieronymus) da Verrazzano, made in 1529. Still another map given by Giovanni da Verrazzano to Henry VIII but no longer extant was known to Hakluyt, who in 1582 published a map by Michael Lok that made use of it. It is clear, therefore, that the planners of the Ralegh voyages were acquainted with Verrazzano's exploration.

On January 17, 1524, Verrazzano and his crew set sail due west from the Madeira Islands at 32 degrees north latitude. Driven off their course, as Verrazzano wrote, by "a storm as violent as sailing man ever encountered,"[7] they made landfall about "34 North Latitude" on the low-lying coast south of present Cape Fear. Sailing south for about "50 leagues" (Verrazzano's sea distances are usually generalized estimates, in contrast to his land-based observations of latitude, made with astrolabe or quadrant), he returned to his landfall because he found no har-

5

bor or strait to the south. He landed, explored the country with great delight in its beauty (it was early March), and met the natives, who at first were uniformly fearful, then friendly. One encounter, amusing but initially terrifying, occurred when Verrazzano sent a small boat ashore to obtain needed water. The "coast was so churned up by enormous waves" that the boat could not land; a young sailor, attempting to swim, was rescued from drowning by the natives, who carried him up the beach by his arms and legs to a fire, at which they apparently intended to warm and revive him. Thinking he was about to be roasted and eaten, the sailor shrieked and struggled desperately. His offshore companions looked on in horror until the Indians made gestures of friendship, withdrew to a sand dune, and allowed him to swim back to the boat.[8]

This episode occurred while Verrazzano and his men were sailing east along the North Carolina barrier islands between Cape Fear and Cape Lookout. After rounding Cape Lookout they sailed along the Outer Banks of North Carolina, where Verrazzano made a geographical misinterpretation that profoundly influenced the beliefs and actions of many explorers for the remainder of the sixteenth century and later. This misinterpretation is best described in Verrazzano's own words, which are found in the Cèllere Codex copy of a letter he wrote to Francis I upon his return to France:

> We called it [the country around Cape Lookout] "Annunciata" from the day of arrival, and found there an isthmus one mile wide and about two hundred miles long, in which we could see the eastern sea from the ship, halfway between east and north. This is doubtless the one which goes around the tip of India, China, and Cathay. We sailed along this isthmus, hoping all the time to find some strait or real promontory where the land might end to the north, and we could reach those blessed shores of Cathay. This isthmus was named by the discoverer "Varazanio," just as all the land we found was called "Francesca" after our Francis.[9] [The length of the Outer Banks from Cape Lookout to Old Currituck Inlet is actually 190 miles.]

Both Vesconte de Maggiolo's map (1527) and Gerolamo da Verrazzano's chart (1529) show the isthmus; beside it, Gerolamo shows, offshore in the Atlantic, this legend: "From this eastern sea is seen the western sea. 6 miles from one to the other."[10]

Thus Verrazzano described the Outer Banks as an isthmus, and Pamlico and other sounds, as seen from the masthead of the *Dauphine,* as the Pacific Ocean leading to Asia. It was not the first time or the last that an explorer interpreted what he saw as what he desperately wanted or expected to see. As will be seen, the hope of finding Verrazzano's bay or sea lured expeditions westward in Carolina and Virginia for at least a century and a half.

That Giovanni da Verrazzano did not see any inlet as he sailed along his "isthmus . . . about two hundred miles long . . ." is not strange; a good navigator sailing along a stormy coast is obliged to avoid shoals, spits, rocks, treacherous currents, and sudden unfavorable winds. Verrazzano encountered all of these along the coast of North Carolina, and he apparently stayed too far away from the shore to spot inlets; that he survived to return safely is a tribute to his vigilant caution, as well as to his seamanship.

Beyond "Annunciata," Verrazzano and his men followed "the coast which veered somewhat to the north [at Cape Hatteras?] and after fifty leagues reached another land . . . full of great forests. We anchored there, and with XX men penetrated about two leagues [five or six miles?] inland, to find that the people had fled inland in terror into the forests."[11] Following three days of exploration, during which Verrazzano and his men found settlements but no Indians, they came upon an old woman hiding in the grass with some children. They took from her an eight-year-old boy, whom they carried back to France. They sailed along the coast,

which we baptized "Arcadia" on account of the beauty of the trees. In Arcadia we found a man who came to the shore to see who we were; he stood suspiciously ready for flight. He watched us but would not come near. He was handsome, naked, with olive-colored skin, his hair fastened back in a knot. There were about XX of us ashore, and as we coaxed him, he approached within about two fathoms of us, and showed us a burning stick, as if to offer us fire. And we made fire with powder and flint, and he trembled all over with fear as we fired a shot. He remained as if thunderstruck, and prayed, worshiping like a monk.[12]

The Europeans apparently were ignorant that the Indian, as historian and biographer L. C. Wroth suggests, was offering them a peace pipe. He was almost certainly offering them not just a burning stick but a tobacco-reed pipe of friendship, a ritual about which the foreigners could not have known, even if they had heard of the smoking of tobacco. These incidents in "Arcadia" could have occurred almost anywhere between Cape Lookout and the Delaware River, as here Verrazzano is not precise geographically or chronologically. The area of present Kitty Hawk; the mainland between old Currituck Inlet and Cape Henry, south of the entrance to Chesapeake Bay (which Verrazzano apparently sailed by without seeing); and the Virginia-Maryland peninsula have all been suggested as the site of "Arcadia."[13] Each has its difficulties. More important than the exact location is the impression one receives of these Indians from Verrazzano's vivid description of their appearance, manners, and reactions to the Frenchmen. Wherever along the coast the recorded incidents took place, the strong impression given to Elizabethan readers of Ramusio's or Hakluyt's printed version of Verrazzano's report must have been of cautious but kindly and friendly natives inhabiting the area they were planning to colonize.

From "Arcadia" Verrazzano continued northward along the coast, entering present New York harbor and the Hudson River ("Angoulême") and Narragansett Bay and Newport harbor ("Refugio") before rounding Cape Cod ("Armeline Sirtes") and turning east from Cape Breton Island for France and Dieppe harbor. There, on July 8, 1524, he wrote his report to the king, six months after his departure from the Madeira Islands at the beginning of 1524.

It was the maps that recorded Verrazzano's voyage and showed the narrow isthmus that separated the Atlantic from the Pacific, however, and not the written report, that excited and influenced the hopes and plans of later explorers. The letter to the king, as eventually published from an abbreviated manuscript by Ramusio in 1556 and translated into English by Hakluyt in 1582, lacks any reference to the isthmus or to numerous other items or place names found on the maps; the Cèllere Codex, quoted above, remained in private libraries for almost four centuries until found and published in 1909.

The two great world maps previously mentioned—Vesconte de Maggiolo, 1527, and Gerolamo da Verrazzano, 1529—are interesting and instructive examples of different delineations of a coast explored for the first time—delineations as interpreted by two skilled and able cartographers who based their work on similar reports. Both of these men were influenced by the necessity of incorporating their new information into the framework of a world map, but the maps they produced reflected differing interpretations of previous reports and existing maps. They were obliged to adjust their work to a coastline between previous explorations to the south (Hernando de Soto's discovery of the Florida peninsula, shown by the heavily marked coastline in Verrazzano's map) and to the north (the Cape Breton Island-Newfoundland-Labrador area), with its complicated geography given contradictory interpretations from the reports of English (Cabot), Portuguese (Corte Reals and José Alvarez Fagundes), and French explorers.

It is clear that both Vesconte de Maggiolo and Gerolamo da Verrazzano had access to Giovanni da Verrazzano's maps or coastal sketches and probably the ship's log, which was sufficiently detailed to make possible unequivocal identifications of many landmarks (New York harbor, Narragansett Bay, Cape Cod, etc.). The toponymy on both planispheres is much fuller than that in the texts of the letters that have survived. Both maps agree with the Cèllere Codex concerning many place names. In addition, each exhibits nomenclature found on neither of the other two documents, as well as names found only on the two maps. These names occur at times in a different order and, to add to the confusion, are repeated on the maps.[14] In this regard Verrazzano was more cautious than Maggiolo, but the Verrazzano map includes more names. As an executor of Giovanni's estate following his death in 1528, Gerolamo must have had full access to his brother's papers in the construction of the 1529 chart. Gerolamo's other Verrazzano-type maps will be discussed later.

In his later maps, Maggiolo turned to the Spanish type of coasting delineation. He was a skilled and highly regarded map maker, but, as Italian cartographer Giuseppe Caraci demonstrated in a study of his methods, his work was characterized by "oscillations and contrasts." Maggiolo often reverted to old, out-of-date delineations rather than showing a steady progression of new information. It was not until his last dated atlas, December 10, 1549, that Maggiolo returned to Giovanni de Verrazzano's discovery. In a planisphere he portrayed vividly the myth of the sea "Mare Indicum" without any coastal toponymy.[15]

In England, Verrazzano's conception of a narrow isthmus did not depend on these two maps; for some reason or purpose no longer known, the explorer sent a map and a globe to Henry VIII about 1525 or 1526. In his *Diuers voyages* of 1582, Richard Hakluyt, in enumerating his reasons for his belief in a northwest passage, wrote: "Secondly, that master Ion Verarzanus, which hath been thrise on that coast, in an old excellent mappe, which he gave to king Henrie the eight, and is yet in the custody of Master Locke, doth so lay it out, as it is to bee seene in the mappe annexed to the end of this boke, being made according to Verarzanus plat."[16]

This map, no longer extant, was apparently an earlier version of Gerolamo da Verrazzano's 1529 map, presently in the Vatican. For Hakluyt's *Diuers voyages*, Michael Lok created a map that incorporated Verrazzano's chart ("The mappe is master Lockes") with the isthmus at 40 degrees north latitude (see plate 21). This is similar to the 41 degrees north latitude shown on Gerolamo's chart rather than the 33 degrees north latitude shown on the Maggiolo 1527 map and its derivatives.[17] Unfortunately, the Lok engraved map is not a copy of Verrazzano's chart; rather, it is apparently an attempt by Lok or Hakluyt to include the other major contemporary theories concerning the possible existence of northwest passages to the Orient through or around the North American continent. Hakluyt believed that such theories were important possibilities to be considered as part of the colonization projects, for which he was so influential an adviser.

Hakluyt's continuing interest in the Verrazzano sea revealed itself in references to the map and to a globe in his *Discourse of Western Planting*, which he wrote in 1584, the year of the Amadas and Barlowe voyage. He particularly supported the theory of a passage to Cathay at this point, describing the map more fully as a "mightie large old mappe in parchment made as yt should seems by Verarsanus, traced all along the coaste from fflorida to Cape Briton with many Italian names, wch laieth oute the sea making a little necke of land in 40 degrees of latitude much like the streyte neck *or* Isthmus of Dariena." He later commented on "an olde excellent globe in the Queenes prive gallery at Westminster wch. also seemeth to be of Verarsanus making." This globe, he wrote, shows "the very selfe same streite necke of lande in the latitude of 40° with the sea joining harde on both sides as it dothe on Panama and Nombre di Diós."[18]

Knowledge of the Verrazzano sea and isthmus among the planners of Elizabethan transatlantic colonization was not limited to Hakluyt's references and the Lok maps, however.[19] One of the most popular maps of the New World throughout the sixteenth century was Sebastian Münster's "Novae Insulae," a woodcut designed by the German cosmographer in the late 1530s and first published

9

in the Ptolemy *Geographia* of 1540 and later in Münster's own *Cosmographia*. This map, reprinted in six editions of Ptolemy, 1540-1552, and in thirty-five editions of the *Cosmographia Universalis* in Latin, German, and Italian until 1628, divides the New World into three continents: "Francisca," "Terra florida," and the "Atlantic Island that they call Brasil or America." The great bight or bay dividing the two northern landmasses is unnamed; a wide isthmus separates it from the Atlantic. In the same volume appeared another double-page map—a world chart, with the Verrazzano sea and isthmus.[20]

Among the numerous manuscript maps from the sixteenth century that are extant, none is more certain of being known to the planners, and perhaps the navigators, of the Roanoke voyages than a polar projection chart drawn by Dr. John Dee for Humphrey Gilbert in 1582 in preparation for the latter's voyage the following year (from which he did not return). Dr. Dee, a learned mathematician, cartographer, alchemist, occasional charlatan, and favorite astrologer, physician, and adviser to Queen Elizabeth, was, like Gilbert, an ardent proponent of a search for a westward passage. Gilbert, lost with the sinking of his vessel while returning from his 1582 voyage, for which this chart was made, was the half brother of Walter Ralegh. Upon Ralegh devolved the earlier grant to Gilbert of an extensive part of North America. Dee evidently demonstrated on this map different possible westward passages to the South Sea, Japan, and Cathay that were surmised but not probed by explorers. The map shows the great open Northwest Passage north of Labrador and "Estotiland"; a "rio s. pero" that is a prolongation directly west of the St. Lawrence River; the "Mare de Verazana 1524" at 38½ degrees north latitude, north of the Spanish "Arena" (Cape Cod) and St. Mary's Bay (35 degrees north latitude); and a river at 32 degrees north (St. Elena?) that leads to a large lake with an outlet into the Gulf of Mexico. A strait connects the St. Lawrence River to Verrazzano's sea. Both Dr. Dee and Hakluyt were also compilers, or copiers, of additional maps that included no such westward passages; they evidently maintained open or skeptical minds.

III. The Ayllón-Gómez Voyages

European cartography of the east coast of North America after 1530 was dominated for the remainder of the century not by the more precise and hydrographically correct record of Giovanni da Verrazzano (except for his isthmus-sea misconception) but by the coastal morphology and place names resulting from the Spanish expeditions of the Licenciado Lucas Vasquez de Ayllón and Estéban Gómez. It is the contention of this study that Elizabethan explorers, and specifically the Roanoke colonists, were strongly influenced in their conception of the North American coast and in their actions by this Spanish cartography. To support this contention, however, requires a somewhat intricate study of the Spanish voyages, the resultant maps, the Spanish system of storing and secreting information about discoveries, and the ways in which this information became available to the Elizabethans.

Ayllón was the first name given by Europeans specifically to the area that is now Carolina.[21] The Licenciado Lucas Vasquez Ayllón, a wealthy auditor and influential judge of Santo Domingo, prepared an expedition to search out new lands that he had heard lay north of Florida, which Ponce de Leon was attempting unsuccessfully to colonize.[22] At the end of 1520 Ayllón dispatched a caravel under the captaincy of Francisco Gordillo with specific orders to befriend and not enslave any Indians encountered. In the Bahamas Gordillo met another Spanish caravel under the command of Pedro de Quexós that had previously been sent out by another auditor, Juan Ortiz de Matienzo, in search of Indians to capture as slaves. Previous raids had stripped the islands of their population; the notoriously inhumane treatment of the Indians, in mines and fields, by their owners, who regarded the natives as expendable, had rapidly reduced the supply. Quexós, facing an empty-handed return to Santo Domingo, agreed to join Gordillo; together they sailed northwest, making landfall in June, 1521, at north latitude 33 degrees 30 minutes. There they entered a large river, which they named Rio de San Juan Batista (River of St. John the Baptist), and formally took possession of the surrounding land for the king of Spain.[23] Disregarding Ayllón's express command, they enticed 150 natives aboard the two caravels and sailed back to Santo Domingo to sell them. Misfortune followed. One of the ships sank, and Diego Columbus, who headed a royal commission, ordered the Indians freed and returned to their homeland.

Ayllón then sailed for Spain, where he received in June, 1521, a cédula, or royal grant, of the territory and the titles adelantado and governor. The king also stipulated that Ayllón explore the coast for 800 leagues, searching for and investigating any westward passage that he might discover. Delayed in preparing for colonization, Ayllón sent Quexós on an additional exploratory mission in 1525. Quexós probed the North American coast for 250 leagues, probably from Florida to about 38 degrees north latitude (just north of Chesapeake Bay).

It was not until July, 1526, that Ayllón's party landed at the mouth of a large waterway that they named the River Jordan. Fernandes Oviedo, a careful historian who based his statements on interviews with Quexós, the expedition's pilot, three Dominican friars, and others, wrote that landfall was at north latitude 33 degrees 40 minutes, which would have been about ten miles north of the site of Quexós's first landing in 1521.[24] Members of Ayllón's expedition brought five well-equipped ships with more than 500 colonists. Upon entering the river, they lost the flagship and all its supplies in a wreck. Among the colonists were twenty black slaves, the first forerunners of the sad sequence of more than 300 years. To Ayllón, however, must be given the credit for also bringing the notable Father Antonio de Montesinos, the first person to denounce Indian slavery in the West Indies, and two fellow Dominican friars.[25]

The colonists did not find the location suitable, and, in the words of Oviedo, "they decided to go and settle on the coast beyond, in the direction of the west coast; and they went to a large river (40 or 50 leagues from that place, more or less) called Gualdape; and there they established their camp or settlement on that coast." Evidently the colonists followed the coast to the southwest. The name of the town they built was San Miguel de Gualdape. "They began to build their houses," Oviedo continued. "The land was level and very marshy but the river was powerful. . . . At the entrance of it was a bay which ships could enter only at high tide."[26] No extant map includes the name Gualdape (or San Miguel de Gualdape), and neither the identity of the colonists' point of departure nor the length of their league (potentially from 1½ to 3½ nautical miles) is certain.[27] There are vigorous advocates for places from the Altamaha in Georgia to the James River in the Chesapeake; Winyah Bay (at the southwest end of the great inward curve of the Atlantic coast from the Cape Fear River at 33 degrees 52 minutes north latitude nearest to the stated landfall of 33 degrees 40 minutes) and the Savannah River are probably the two most frequently defended, although Santa Elena (Port Royal) has increasing support.[28]

Pertinent because they date from only fourteen years after the brief settlement at San Miguel de Gualdape by Spanish explorers are the comments by Luis Hernandez de Biedma, factor of the Hernando de Soto expedition, and the "Gentleman of Elvas," a pseudonym, possibly for Alvaro Fernándes, who kept a journal of the expedition. In 1540 Biedma wrote that the members of the de Soto expedition reached "the town of Cofitachique . . . seated on the banks of a river which we believe to be the Santa Elena, where the Licentiate Ayllón had been." The members of the expedition opened graves in the local temple and found Castilian items: axes, a rosary, false pearls, "all of which we suppose they got in exchange, made with those who followed the Licentiate Ayllón. From the information given by the Indians, the sea should be about thirty leagues. We knew that the people who came with Ayllón scarcely entered the country at all; that they remained continually on the coast, until his sickness and death. . . . One whose lot it was to have been among them told us that of six hundred men who landed only fifty-seven escaped."[29] The Gentleman of Elvas, writing about the Spanish objects found at Cofitachique, said that "the Christians, the Indians said, had many years before been in the port, distant two days' journey. He that had been there was the Governor Licentiate Ayllón."[30]

If the identity of Cofitachique as Silver Bluff, 13 miles southeast of Augusta, Georgia, on the Savannah River, is accepted, the location of Ayllón's settlement as being at the mouth of the Savannah does not necessarily follow. The information given by Biedma and the Gentleman of Elvas is evidently inexact and contradictory: the language barrier was difficult, the distances uncertain, and the belief that the Savannah was the Rio Santa Elena wrong. Actually, Winyah Bay, apparently located by some contemporaries as the place of Ayllón's settlement, is only 30 miles farther (160 miles) in a straight line from Silver Bluff than is Savannah Sound or Santa Elena-Port Royal (130 miles).

Misfortune followed upon misfortune. The loss of the main supply ship soon resulted in a lack of food and later starvation; the swampy ground contributed to, if it did not cause, sickness and fever; the cold became intense. Ayllón himself became sick and died on October 16, 1526. Malcontents mutinied and were in turn overthrown and executed. The colonists decided to return to Santo Domingo. The winter must have been exceptionally severe; seven men froze to death during the return voyage. Only 150 of the original 500 to 600 men reached Santo Domingo alive.

The first known map to record the explorations of Ayllón and to establish a new type of nomenclature for the southeastern Atlantic coast is a large mappemonde by Juan Vespucci, nephew of Amerigo Vespucci, the Italian merchant-adventurer for whom America was named. The map, made in 1526 and presently in the possession of the Hispanic Society of America in New York City (see plate 4), was probably drawn in the Casa de Contratación, the Hydrographic Office of the Board of Trade, although it was not a copy of the official world map. Juan Vespucci was appointed pilot major of the Casa de Contratación in May, 1512; in 1515 he was a member of the junta that was brought together to improve, correct, and pass upon the padrón réal—an official royal chart on which all discoveries made by returning pilots, reported and attested by oath, were to be entered and corrections made—and to oversee the issuance of charts to outgoing navigators. In 1516 Vespucci was appointed one of the pilots to the king of Spain. By 1524 he became an adviser to the Badajoz Commission, convened (unsuccessfully) to resolve a territorial dispute

13

between Spain and Portugal over ownership of the Moluccas. In April, 1526, he was appointed to examine pilots during the absence of Sebastian Cabot, royal pilot major.[31]

It was at this time that Vespucci produced his famous mappemonde, which was evidently based on preliminary information reported by Quexós from his second expedition of 1525. The map does not include the names Rio de San Juan Batista, the sole landfall of the first voyage, or San Miguel de Gualdape, from Ayllón's colonizing voyage. No information from this last voyage could have reached the Casa de Contratación in time for inclusion on Vespucci's map.[32]

One name that is still in use appears on Juan Vespucci's map for the first time: "C. de Sāta elena" is present Hilton Head, at the entrance to Baya de Santa Elena (present Port Royal).[33] On this map appears, also for the first time, "C. de sā mā" and "baya de sā mā" (the Cape and bay of St. Mary). The distance from the Cape of Trafalgar (Cape Lookout), with two intervening rivers and the narrow mouth of the bay, indicates that Quexós discovered and entered Chesapeake Bay, which is depicted on Vespucci's map as round in configuration. The entire coastline north of Florida to the limits of Quexós's exploration, shortly beyond the Bay of St. Mary, lies east-west in the same latitude. The abrupt end of the coastline, with an open gap to "Bacalaos" (here Nova Scotia and Newfoundland), demonstrates that Vespucci was meticulous in recording only information actually reported to him.[34] Yet, the voyage of Estéban Gómez along the North Atlantic coast had already been made and had been mapped the previous year by another cartographer of the Casa de Contratación.

Gómez, the able Portuguese pilot and captain, believed a more direct passage to Cathay and the Spice Islands than Magellan's long route around South America lay in the North Atlantic between Florida and Labrador. He persuaded Charles V of Spain to equip a vessel under his command to find the way. He left from the northeastern port of Coruña on September 25, 1524, two and a half months after Verrazzano had returned to Dieppe and made his report, and half a year before Quexós left Santo Domingo on his second voyage. Surviving details of the Gómez expedition are chiefly cartographic. But Antonio Galvão, a contemporary of Gómez and a fellow native of Oporto, wrote: "He [Gómez] went from Galicia to the island of Cuba and to the point of Florida, sailing by day because he knew not the land and because he wished to visit every bay, harbor, or inlet, to see if he could reach the other coast."[35] Gómez returned to Coruña on August 21, 1525, eleven months after his departure. Frustrated by the failure of his mission, he seized and brought back a shipload of Indians to be sold as slaves. Most of his time was spent in a careful search of the New England coast between Cape Cod and Penobscot Bay; there is little evidence of exploration along the southeastern portion of the North American coast.

14

Before leaving Spain, Gómez must have learned of the royal cédula of June 12, 1523, which granted Ayllón settlement and possession of the present Carolina region, with express royal authority to search for a westward passage to the Spice Islands. If Gómez sailed by the trade winds to the West Indies, he would there have learned of Ayllón's intensive preparations for the settlement and for the exploratory second voyage of Quexós, which was undertaken in the spring of 1525.

It is the maps made from reports of these voyages that evolved into the navigational charts that were used by the Roanoke colonists and that influenced their decisions. The earliest map to show the discoveries of Gómez's voyage is the Castiglioni world chart of 1525. It is the first of a series of maps by or attributed to Diego Ribero (Portuguese spelling: Diogo Ribeiro), a Portuguese pilot, maker of navigational instruments, and the finest chart maker in the service of Spain, that established the general contour and toponymy of the east coast of North America in Spanish maps for more than a century. It is also the earliest dated map to show an uninterrupted coastline from the Strait of Magellan to Labrador based on actual discovery. The Castiglioni world chart was commissioned by Charles V for his friend Count Baldassare Castiglioni of Mantua, envoy of Pope Clement VII, in whose family it has remained.

Ribero, who was in Coruña at the time of Gómez's return, had recently arrived there to take up his duties in the Casa de la Especieras, the Spanish office of the spice trade, where he was the official cartographer. He may have brought the vellum sheets with him from Seville almost complete and copied from sheets in the hydrographic office there, inasmuch as the coasts are heavily drawn in green, except for one lightly sketched in black between Florida and Cape Breton.[36] This portion of the map was drawn from the information provided by Gómez and his crew. There is no evidence that Gómez touched the coast more than casually before reaching New England. From Florida to Cape Cod a gently inverted S is drawn, without a place name or identifying landmark such as Cape Fear, Cape Lookout, or a bay or islands. The map's cartographical importance is that it established for all Spanish-type maps for more than a century a curving, generally northeastward shoreline, in contrast to the incorrect Vespucci coast shown aligned directly east to west. Along the New England shore, where Gómez spent months in careful but unsuccessful exploration for a westward passage before returning, are six place-names between Cape Cod and Penobscot Bay, two of the most easily identifiable landmarks on the east coast but here unnamed.[37]

Meanwhile, in 1521 reforms in the geographic-hydrographic department of the Casa de Contratación had resulted from a visit by inspectors to Seville on orders from Charles V.[38] The Casa had been established in 1503 with extensive and powerful authority over foreign trade and affairs. In 1508 Amerigo Vespucci (who died in 1512) was appointed pilot major; along with a cosmographer major, he was authorized to examine and license pilots, to inspect for accuracy all navigational instruments, and to keep a padrón réal. In 1526, however, Charles V found no padrón réal but only charts that were contradictory and erroneous. Juan Vespucci, who had succeeded Amerigo Vespucci as pilot major in 1512, fell out of royal favor about this time, although the reason why is not certain. Charles V ordered that the new pilot major, Sebastian Cabot, construct a padrón (thereafter called padrón general). But Cabot had left two months earlier, in August, on a voyage of discovery to South America that was to last for four years; consequently, Ferdinand Columbus (the illegitimate but highly ranked son of Christopher Columbus) was ordered to make the padrón general with Diego Ribero as his assistant.[39] Ribero did not reach Seville from his duties in Coruña until the summer of 1528. Columbus appointed Alonso de Chaves as his assistant, and Chaves became cosmographer upon Ribero's arrival. In 1530 Alonso de Santa Cruz returned with Sebastian Cabot, was appointed cosmographer major, and occupied that position with many additional honors until his death in 1567. The Casa attracted to Seville additional expert cosmographers and pilots, not only from Spain but also from Portugal, Italy, and elsewhere. The hydrographic office in Seville became unrivaled by any other such government organization, past or contemporary.

Spanish pilots were required to use only the navigational charts copied from the padrón réal (or general) or approved by the pilot major.[40] The value of this procedure as a means of minimizing the number of shipwrecks in unknown seas is evident, as is its usefulness in improving cartographic knowledge. The master chart kept in Seville had to be large in order to accommodate worldwide details and corrections.[41] No original padrón has survived, nor even a copy certified by the pilot major.[42] Maps that have survived from this period are predominantly those with illustrative or artistic appeal rather than ones made for navigational use.

By 1529 Diego Ribero and his associates in the Casa had gathered much scattered information and organized it cartographically. Concerning the land of Ayllón, the historian Oviedo, who had been a friend of Ayllón and his family in Spain, wrote in his *Historia general y natural de las Indias* that he had interviewed survivors of the unfortunate third expedition: Quexós, who had been the pilot on all three voyages and led the exploratory second expedition; Friar Antonio Montesinos; and Antonio Gómez, a lieutenant on the voyage.[43] These and other survivors were available to Ribero and must have been questioned closely by him. Two important maps of this period by Ribero, both dated 1529 and signed by him, have survived: one in the Landesbibliothek in Weimar, the other at the Vatican in Rome. Two others are ascribed to Ribero but are unsigned—one in Weimer dated 1527 and one in Wolfenbüttel; if these maps were not made by Ribero, they were made under his direction.[44]

The Weimar 1529 map signed by Ribero marks a great advance in knowledge over the unsigned Weimar 1527 map, which may be a copy of a map made by Ribero before he returned to Seville. The 1527 map has many of the decorative

16

characteristics of Ribero's signed maps and has as well the basic coastal contours that Ribero established; but, as stated previously, it is unsigned but dated as having been "made . . . at Seville."[45] The Weimar 1529 map also has many place-names that appear for the first time—names that became the established nomenclature and replaced a few of the Vespucci 1526 names.

The Weimar 1529 map is a great cartographic achievement, but it does not equal the Ribero 1529 map in the Vatican, which is a masterpiece in artistic composition and illustration, in sharpness and skill of detail and of lettering, and in the inclusion of drawings of navigational instruments such as the astrolabe. The Vatican map has more names on the North American coast than does the Weimar map, and the legends describing the territories of Gómez and Ayllón are longer. The Wolfenbüttel map, of the Western Hemisphere only, is undated and unsigned, but it is generally ascribed to Ribero because of similarities of style, contours, and place names. Professor L. A. Vigneras of George Washington University has demonstrated that it was probably made shortly before October 20, 1530, since it refers to the imminent departure of Diego Ordaz, in command of a Spanish naval expedition, from the Spanish port of San Lucar.[46] There are more place-names on the east Atlantic coast than on any of the previous maps, but it is sparser in decorative additions. South American legends are full and numerous, but North American legends are fewer and shorter ("Tiera del Licenciado Ayllom," with its Portuguese spelling of Ayllón, has none).[47]

Whether revisions of the padrón general lapsed after Ribero's death in 1533 or Charles V's order was never followed is not clear; but in Charles's absence in 1535 Isabella of Portugal, his queen and regent, ordered Ferdinand Columbus to see that the order be completed "with all possible despatch." The command was effective, for Alonso de Chaves, the pilot major, had constructed a model sometime during 1536. Oviedo states that he received a copy of the Chaves padrón general in that year,[48] although no authenticated copy of it exists.

It is possible, however, to learn in detail the information included in the padrón general of 1536 because of a large unpublished work (1538-1539) on navigational methods and instruments made by Alonso de Chaves that, in the fourth book, has a carefully made *derrotero,* a rutter or navigational guide, that lists the coastlines of the New World with the nomenclature of settlements, rivers, capes, bays, and the direction and distance of each to the next.[49] This manuscript, completed by 1539, is valuable for its summary of the information gathered from many sources (such as the survivors of the Ayllón settlement) and interpreted by the cartographers of the Casa.

No extant map, however, follows it exactly, nor does it solve many problems of identification of names on Spanish or "Sevillian"-type maps of the sixteenth century, with modern nomenclature. Until the end of the century most European maps of the New World followed the pattern established by Ribero, Chaves the pilot major, and Santa Cruz the cosmographer major. While Spanish-type maps may retain the same general coastal contours, they exhibit confusing variations, additions, omissions, and changes in names and landmarks, as well as in their order and sequence along the coast. An examination of the details on the maps from the original padrón reveals some of the errors and mutations that must have been misleading to navigators of the period who voyaged along the coast, as well as to the planners of the Roanoke colonies and the navigators of those voyages.

The two most prominent projections on the east Atlantic coast between the Cape of Florida and Newfoundland are Cape Lookout and Cape Cod: Cape Lookout is Cabo de Trafalgar, and Cape Cod is Cabo delas Arenas. (Monomoy Point, the long, sandy spit that extends south from the elbow of Cape Cod, is Arenas on some early charts; Cabo de Santiago is Cape Cod's northern tip on such charts, though the names Arenas and Santiago are frequently omitted or reversed on other maps.)[50] Between these two points—Trafalgar and Arenas—the topography varies and the toponymy is uncertain. On all Sevillian-type maps the distance between these capes has been telescoped and the coastline between incorrectly portrayed as a single outward or bulging curve [⊃]. Evidently it was a stretch of coastline unseen or barely touched by either Quexós or Gómez. Chaves gave the latitude of Trafalgar (Cape Lookout) as 35½ degrees north, one degree too high, and Cabo delas Arenas (Monomoy Point) as 38⅓ degrees, 3 degrees too low. He stated the distance between them as 62 leagues (approximately 210 miles); the actual distance is not less than 675 miles.

The Chaves padrón general of 1536 contributed two changes or additions to the Ribero-type charts. Within the Baya de Santa Maria, to the north, is a "Rio Salado" [Salt River]; and Cape Cod does not extend northeastward into the sea like a finger. The shoreline turns directly west from its tip, like a 7, instead of having a V-like shape for Cape Cod Bay. A possible explanation of Rio Salado is that it represents Chesapeake Bay, and that Quexós, missing the eleven-mile strip of coast between it and the sounds, regarded both bodies of water as forming one large bay. Later maps exhibit many variations in the delineation of the bay, including a narrower entrance and two rivers—the Rio Espiritu Santo to the south and Rio Salado to the north.[51] Governor Ralph Lane of the Roanoke colony evidently thought that the Roanoke River was the Rio Salado, as will be shown.

For more than thirty years after the Gómez-Ayllón voyages, the east coast of North America attracted little additional attention, except by the French under Jacques Cartier and his successors, who probed the St. Lawrence River, and by the fishing fleets of many nations, which continued to utilize the rich fishing grounds off Newfoundland. Convinced that no Aztec or Inca empires rich in treasure awaited conquest and that the land featured only gold-free soil and primitive Indian tribes, the Spanish and other seafaring European nations turned their energies elsewhere. In the tropics, they believed, were the realms of wealth and unexplored possibilities.

Yet, to contemporaries remained the unextinguished hope—even the probability—that somewhere along that northern coast lay an undiscovered strait or westward passage to the riches of the Far East.

The Spanish and Portuguese governments grew increasingly zealous during the sixteenth century in keeping unavailable to other European nations any maps showing useful geographic and navigational details. In 1527 Charles V decreed that no foreigners could be employed by the Casa de Contratación, although those already in its service, such as Sebastian Cabot and Diego Ribero, were not dismissed. Portugal treated its charts as secret state papers; its pilots were forbidden on pain of death from giving or selling charts to foreigners.[52] During the sixteenth century practically no maps or charts were printed in the Iberian peninsula, except for a few general and undetailed examples.

IV. Foreign Maps
Potentially Available in England

France, and indirectly England, drew heavily on Portuguese exiles for the notable school of map makers and hydrographers that developed in the seaport town of Dieppe, the home of the great sea captain Pierre Crignon and of Jean Ango, the powerful merchant and armateur, or type of privateer. The charts and atlases made there, with their exquisite miniature scenes and figures of animals and natives, were influenced by or made by Portuguese artists and cartographers, as shown by occasional Portuguese nomenclature or spelling. All of these, as Dr. Helen Wallis, former map librarian at the British Museum, has demonstrated, drew upon a common body of material—maps, charts, and graphic documents as well as oral interviews and firsthand accounts—that has disappeared.[53] By fitting together the folio pages of the atlases from the Dieppe school, such as the one of John Rotz (1542), the Hague Atlas (ca. 1545), Nicolas Vallard (1547), Pierre Desceliers (ca. 1555?), and Guillaume le Testu (1555-1556), one may suppose, if not deduce, the existence of a great world planisphere about 13 feet in width that was probably in use as early as 1536.[54]

From these and other sources there came to sixteenth-century England maps and atlases that found their way into royal or private libraries and were available to English planners of overseas enterprises such as Dr. John Dee, John Frobisher, Sir Humphrey Gilbert, Sir Walter Ralegh, the two Richard Hakluyts, Sir Francis Walsingham, and others. An unknown number of these maps have been destroyed or have disappeared; others are known to have existed on the basis of contemporary references to them, such as the inventory of Henry VIII's palaces; and a few have survived and are presently available for study.

The disappearance of some maps is especially tantalizing. Where are those once in the extensive library of Sir Walter Ralegh himself? Ralegh was not only a careful student of maps but he also collected them and developed considerable skill in drawing them, as surviving copies of his maps of Guiana attest. Dr. R. A. Skelton, late superintendent of the Map Room in the British Museum, in his "Ralegh as a Geographer," documents the care with which Ralegh studied available maps in planning his seafaring as well as his military undertakings and notes that Sir Walter provided Richard Hakluyt with Spanish maps.[55] Portions of Ralegh's library, including maps, were seized by the government after his beheading in 1618. Some sea charts may have been sent to the State Papers Office, and some of his books and papers were still in the possession of Carew Ralegh, Sir Walter's brother, in 1652 but were subsequently scattered.[56]

The notable collection of maps in the Queen's Gallery at Whitehall, a royal palace in London, in the 1560s was later dispersed or lost in the fire that destroyed the palace in January, 1698. The diarist John Evelyn, visiting Whitehall on September 2, 1680, saw an "Aboundance of Mapps & Sea-Chards."[57] The most celebrated maps in the Queen's Gallery were those made or left by Sebastian Cabot during his several trips from Spain to England during the decade from 1547 to 1557. These included a 1549 revised version of the 1544 Paris world map, with Cabot's claim of finding a northwest passage at 67½ degrees north latitude, which Humphrey Gilbert reported as having examined with other charts before 1566.[58] Hakluyt referred in 1582 to "all his [Cabot's] owne mappes and discourses" as being "in the hands of a William Worthington who was very willing for them to be published."[59] This was never done, however. In 1688 Samuel Pepys, secretary of the Admiralty and diarist, recorded the receipt of forty-five parcels from James VI for the Admiralty Office, of which Parcel 39 was "Sebastian Gabotts maps."[60] Fortunately, not all of the collections in various royal libraries have been lost; the British Library is rich in maps that it acquired from the Royal Library in St. James after 1757, and especially in King George III's topographical collection, acquired in 1828.[61]

In the British Museum is the magnificent Rotz Atlas, presented to Henry VIII in 1542 by John Rotz of the Dieppe school, a Frenchman of Scots descent, a seaman who had been to South America and probably to Sumatra, a hydrographer, and an expert map maker.[62] Disappointed in royal preferments in France, Rotz was attracted to England; but he returned to France in 1547. His map of the North Atlantic coast derives from a Ribero-type source, combined with numerous differences found also on a number of other maps of the Dieppe school. The east coast of Florida extends due north as far as the Cape of Santa Elena at 39 degrees north latitude (this is present-day Port Royal, which is at 32 degrees 14 minutes), then almost due east to the Bay of Santa Maria at 41 degrees, both coasts overextended to twice their actual length. The Bay of Santa Maria is a long, open bay enclosed by two rows of Ribero-type islands; at the north end of the bay is Rio Salado, the earliest appearance of this Chaves-padrón-type name in a dated atlas or map. From Cape Santa Juan (Henlopen?) a long coast without any names runs due north to a point (Cape Cod) at 46 degrees; then another long coast stretches east to Cape Breton and to Cape Race on Newfoundland. Possibly the Rotz atlas was hidden away in the royal library and not consulted.

Another map, undated (ca. 1547) and unsigned, of the Dieppe school, as Dr. R. A. Skelton suggests, possibly hung in the Queen's Gallery at Whitehall and is presently in the British Museum's collection of books and manuscripts begun by Robert Harley (1661-1724), first earl of Oxford.[63] It is the only identifiable map of the Dieppe school that incorporates Giovanni da Verrazzano's "eastern sea" (plate

12). The coastal names are generally of the padrón-general type, though the nomenclature as usual on the Dieppe maps has been gallicized; some names are incomplete, and others are not found elsewhere. The map's significance lies in the fact that John White's general map, "La Virgenia Pars," shows a narrow strait between the Atlantic Ocean and the (unnamed) Verrazzano's sea similar in location (at White's Port Royal vs. the Harleian's R. de S^e helene) and length; these two are the only extant maps exhibiting that geographic relationship.[64]

Another world map, ca. 1539-1540, presently held by the National Maritime Museum at Greenwich, England, is unique in including Verrazzano's sea with a Ribero-type Atlantic coastline and a mixture of French, Italian, and Spanish padrón names.[65] The isthmus is narrow, with no strait connecting the bodies of water; it is at "arenas," just below "C: de s̄ca elena" (like the Harleian and the White maps) and "punta de luluio" (a name on the 1529 Verrazzano that is some distance below the isthmus). The Spanish padrón-type bay with islands, entirely absent on the earlier Verrazzano maps, shows "san luigi" instead of "B. de S. Maria." This unsigned, undated chart is attributed by Marcel Destombes, an authority on Verrazzano's charts, to Gerolamo Verrazzano on the basis of the similarity of the handwriting on this chart to that which appears on the signed 1529 chart held by the Vatican.[66]

Diogo Homem, a Portuguese map maker from a celebrated family of cartographers, lived during the middle years of the sixteenth century in London, where he made his living producing charts and atlases. He had fled Portugal on a charge of murder in 1544 and was in London at least by 1547. Lopo Homem, probably his father and also a prolific producer of atlases and maps, had access to all the charts of the Casa da India, the Portuguese Foreign Office, in Lisbon as early as 1517.[67] From the wealth of new information Diogo produced, the Casa charts were probably available to him also. The records there were destroyed by fire in the disastrous Lisbon earthquake of 1755. Diogo was in London a dozen years or more; although he had received a royal pardon, he apparently did not return to Portugal, for his later work, such as his fine 1568 atlas, presently in the Sächische Bibliothek, Dresden, is dated from Venice. The earlier 1558 atlas, presently held by the British Library, is magnificently designed with gold leaf and colored vignettes (plate 14); it may have been ordered for Philip II by Queen Mary.[68] The North Atlantic chart, like the one in the 1568 atlas, is important to modern historians for its new information on the early explorations by João A. Fagundes and Jacques Cartier in the vicinity of the Bay of Fundy and the St. Lawrence River,[69] but it gives a delineation of the Carolina coast that does not differ materially from that of the earlier padrón-general type.

The Gutiérrez North and South America 1562 (plate 15) is the largest (36¾ inches by 34 inches) and most detailed map of the New World to be engraved and published up to the time of its appearance, and it must have been known and available in England.[70] It was made by Diego Gutiérrez (1485-1554), royal cosmographer and pilot major of the Casa de Contratación in Seville, who had access to the latest Spanish charts and reports of his time. Since Gutiérrez died in 1554, his son, also named Diego, is sometimes credited as author; a chart of the Atlantic made by Gutiérrez, Sr., in 1550 and presently located in Paris[71] exhibits so many similarities that it is the obvious source of most but not all of the engraved map.[72]

The publication of such a detailed map of the New World by a cosmographer and pilot of Spain was, in light of the official attitude of Spain, unparalleled, although "copies" of the padrón evidently were available in other nations as the result of theft, purchase, or gift. Only two copies of this handsomely, even beautifully engraved map are extant: one in the British Library and one in the Library of Congress. The map was probably published in Flanders; the engraver was Hieronymus Cock of Antwerp, an engraver of maps. Artistry evidently was more important than accuracy: the irregularities of the coastline do not conform to any extant manuscript, and such obviously mangled names as "C. de aruz" (C. de Cruz) and "C. gruedo" (C. grueso) would have been corrected with knowledgeable supervision. Baya de Santa Maria exhibits the usual post-Ribero triangular form, with two rivers and several islands scattered within and outside the entrance to the bay. The Florida coast at B. de S. Elena turns eastward at a right angle. To interested privateers, the island-studded Baya de Santa Maria seemed ideally located for seeking safety, making repairs, and obtaining supplies after pouncing on returning Spanish vessels.

V. Exploratory Voyages and Activities along the Southeastern Coast of America in the Third Quarter of the Sixteenth Century

By 1519 the Spanish had discovered the advantages of a return voyage from the Central American ports and Havana by way of the Gulf Stream.[73] That "mighty river in an ocean moving with the volume of a thousand Niagaras" flows northward between Florida and the Bahamas at about 60 miles a day. By 1542 the Spanish treasure fleets followed the current through the Bahama Channel along the coast to 32 degrees north, the latitude of the Bay of Santa Elena (Port Royal), then turned east-northeastward with the Gulf Stream to 38 or 39 degrees north, passing 50 to 75 miles east of Cape Hatteras and the Outer Banks. From there they sailed east to the Azores and Cadiz.[74] Cyclones and hurricanes occasionally took their toll on the ships, driving vessels to shipwreck along the coast, as in the disastrous losses sustained by the great treasure fleets in 1550 and 1553. Survivors had meager chances of being rescued or returned to their homeland. It has been suggested with justice that reports of gold and other ornaments worn by the Indians of the Southeast often had their source in salvage from ship or person.

Before 1564 no European settlement existed along the entire North American coast from the Rio Grande in Mexico to the St. Lawrence in Canada, except for short-lived attempts such as Ayllón's in 1526 or the French Huguenots' in 1562 at Charlesfort, Port Royal. But by the late 1550s the Spanish government demonstrated a renewed interest in "Florida," a term that included the region north to the Outer Banks and even the Chesapeake. Its policy of claiming territorial and economic monopolies for its vast domains in the New World was being challenged, not only by actual colonization by the French in Canada but also by merchant seamen from other nations who, as the Spanish colonists in the Caribbean found, traded needed or better goods at less cost. French and English, as well as Portuguese, vessels found their way to various harbors. Unfortunately for the Spanish, these ships were also well armed; if they encountered an unarmed vessel or a poorly defended town, they might pillage the vessel and loot the town. The great homeward-bound Spanish silver fleets, with accompanying merchant ships, were convoyed by protecting naval escorts; but even then stragglers could be and were picked off. "Many French and English corsairs were going about at Porto Rico and Hispaniola," Pedro Menéndez de Avilés, adelantado of Florida, wrote to King

Philip II of Spain in October, 1565, "and many Portuguese ships were doing the same thing."[75]

The increasing use by pirates of the coast bordering the Gulf Stream and, even more ominous, the rumors of potential French plans for colonization aroused the Spanish court to action. In 1558 Philip II ordered the viceroy of New Spain, Don Luis Velasco, to plant fortified settlements on the South Atlantic coast. Velasco entrusted the task to Tristán de Luna, who left Mexico with a fleet carrying supplies and a small army in June, 1558. Landing at Pensacola on the Gulf Coast of Florida, the expedition bogged down in a "nightmare of ineffectual activity" and was still there two years later. Luna was replaced by Angel Villafañe, who reached Punta de Santa Elena (Port Royal harbor) with a small fleet in 1561. He explored the coast farther north, but his four vessels were caught in a storm before reaching Trafalgar (Cape Lookout). Of these, the two smaller ships with their crews were lost, and Villafañe, discouraged, returned to Havana.[76] Meanwhile, a supply ship sent from Havana in search of Villafañe had sailed as far north as the Chesapeake, where it apparently took aboard an Indian (who was later baptized Luis de Velasco after the viceroy, his godfather). The ship returned to Havana without finding Villafañe.[77]

Earlier that year, a far more able and energetic commander than Tristán de Luna or Villafañe, Pedro Menéndez de Avilés, general of the treasure fleet, dispatched two sloops north from Havana to search out the coast. Avilés made a report of this expedition to King Philip II, but neither it nor any details of the findings of the expedition has survived.[78] In 1565 Philip II made Menéndez the adelantado of all Florida, the vast but unclearly defined grant previously bestowed upon Ayllón and de Soto.

The king and the new adelantado knew from Spanish intelligence sources in France and England that Jean Ribault, a Huguenot captain of proven diplomatic and maritime skill, had, at the instigation and with the support of the Huguenot Gaspard de Coligny, admiral of France, as well as with the tacit approval of the king and the queen mother, planted a colony on the southeast coast of Florida in 1562.[79] The French colony at Port Royal lasted but a few months, but previous French knowledge of the coast is revealingly confirmed by Ribault's statement in his *The whole and true discouerye of Terra Florida*, translated into English and published in London in 1563, that one of his chief reasons for not continuing his planned exploration up the coast to 40 degrees north latitude (Cape Cod) was that "our master pilotes and some others had before byn at some of those places where we proposed to saile."[80]

In 1564 the French under René de Laudonnière established "La Caroline" (Fort Caroline), a fortified colony on the May (St. John's River); Pedro Menéndez de Aviles reacted with vigor and the full support of Philip II. He reached Florida south of the French colony after Ribault arrived at Fort Caroline with a supporting French fleet and supplies for the colony. Marching overland from a landing place he named

St. Augustine, Menéndez surprised the French at Fort Caroline early on the morning of September 21, 1565, and later massacred the entire colony as Lutherans and pirates, except for some women, children, and Roman Catholics and a few who escaped from the fort in the darkness and reached a small French ship down the river.[81]

Among those few was Jacques le Moyne de Morgues, a Huguenot artist appointed by the king through Laudonnière as painter and map maker for the expedition.[82] Le Moyne may have sent some versions of his drawings back to France aboard an earlier departing ship, inasmuch as it is unlikely that he carried a painter's portfolio with him after leaping out of his hammock at the sound of blows and spending three days and nights in swamps and swimming across streams before being rescued by the supply ship. At any rate, a map of Florida by le Moyne was published in 1591 by Theodor de Bry in the second volume of his popular series of works on America, a series richly illustrated with pictorial engravings and maps.

Cartographically, the French expeditions, short and disastrous though they were, supplemented the work of the Spanish and Portuguese cartographers of the southeastern region between St. Augustine and North Carolina. The contemporary sixteenth-century influence of these expeditions was limited before de Bry's 1591 publication, although le Moyne's work was well known to the planners of the Ralegh expeditions and his drawings and map were used by John White, as will be shown. Both Jean Ribault and his chief pilot, Nicholas Barré, were in England in 1563; their charts of the 1562 voyage have not survived, but a Spanish spy's copy of Barré's detailed chart was sent to Spain by the Spanish ambassador in London. Ribault named the rivers that he saw along the coast after French rivers: Seine, Somme, Loire, Charente, etc. The tracing of Barré's chart, presently in the Museo Naval, Madrid, is sufficiently detailed to permit identification of more than three dozen topographical features through their modern coastal equivalents.[83] Le Moyne, who went to London by 1580 when the position of Huguenots in France deteriorated following the murder of Admiral Gaspard de Coligny in 1572, entered the service of Sir Walter Ralegh, possibly after being introduced to him about 1585 by a fellow painter, John White.

Although le Moyne's personal knowledge probably extended little north of Fort Caroline, his experiences in Florida and his paintings, maps, and commentaries must have been valuable to the planners of the Ralegh voyages. Of his mission in 1564, he wrote: "My precise role [was] to chart the sea-coast and to observe the situation of the towns and the depth and course of the rivers, and also the harbors, the houses of the people, and anything new there might be in the province."[84] The fact that John White used or incorporated into his own work le Moyne's drawings and mapping of the country is a visible indication of the Frenchman's contribution to the Ralegh venture. Le Moyne died in 1587; his drawings and map were purchased by de Bry from his widow in 1588 and published in Frankfort in 1591.

Since the map does not chart territory north of Port Royal, the extent of Ribault's 1562 coastal exploration, it does not pertain directly to the mapping of coastal North Carolina. The information in it, however, was soon combined by Dutch cartographers such as Jodocus Hondius with that found in the map drawn by John White and engraved by Theodor de Bry in 1590, and it influenced the conception of the region directly or indirectly until well into the eighteenth century.[85] Unfortunately, the following geographic misconceptions incorporated by le Moyne in his map were seized upon by later geographers: a great body of water at the top of the map, probably deriving from Verrazzano's arm of the Pacific Ocean near the headwaters of rivers emptying into Port Royal; an almost triangular shape for the Florida peninsula; a mountain range extending down its center; and, in western Carolina, the "Montes Apalatci in which gold, silver and copper are found," as well as a lake "where natives find grains of silver."

The French attempt to plant colonies and outposts in Florida was effectively ended by the destruction of Fort Caroline, although French corsairs, often under the guise of merchant vessels, continued to use harbors along the coast. In 1568 Dominique de Gourgues, who had himself been captured by the Spanish and had served them while chained to the bank of one of their galleys, assembled a small fleet well equipped with soldiers and executed a fearful revenge in an attack on San Mateo, a rebuilt Fort Caroline, and two supporting blockhouses at the mouth of the river. Those not put to the sword were hanged from trees with a sign posted nearby that read "I do this not as to Spaniards, nor as to Marranos but as to traitors, robbers, and murderers." This was in obvious response to the similarly phrased notice posted by Menéndez de Aviles after the 1565 massacre: "I do this not as to Frenchmen but as to Lutherans."[86] Although in the ensuing decades scores of French and English pilots and navigators used and knew the coast, particularly below Cape Lookout, their hydrographic knowledge was not transferred to maps that have survived.

Spanish coastal explorations were undertaken officially and recorded, even if some reports have not survived. Menéndez's energy did not flag with the destruction of the French colony and establishment of a Spanish settlement and fort at St. Augustine. Before returning to Spain in 1567 to report to the king and gather additional settlers and supplies, Menéndez had explored 300 leagues of Florida coast and established three forts for the rescue of shipwrecks. He also engaged in the conversion of the natives, provided defense against "corsairs," and sent deep into the interior expeditions under Captain Juan Pardo from San Felipe, the fort on present Parris Island at Santa Elena. The adelantado believed that the silver mines of New Mexico were only a few days' march beyond the mountains.

The geographic conceptions of North America held by Menéndez included other basic errors. One of these was a belief that from the Baya de Santa Maria there might be a strategic passageway to China as well as to the silver mines of San Martin and Zacatecas in Mexico. Menéndez had heard a version of this story as early as 1554 from a Biscayan sailor who had told him of sailing with a French pirate through an opening at about 45 degrees north latitude (the Gulf of St. Lawrence) to a large bay (Indian accounts of the Great Lakes?) at 42 degrees north. This bay was separated from an arm of the South Sea (a Verrazzano-like intrusion from the Pacific) by an isthmus only a quarter of a league in width. The sailor also told Menéndez that the Rio Salado was a waterway lying between Santa Maria and this bay, in effect forming a great continental island south of the St. Lawrence and east of the Rio Salado that included present New England and New York State.[87]

Menéndez wrote to the king on October 15, 1565, a few weeks after the extermination of the French at Fort Caroline, that he planned to establish a fort at Baya de Santa Maria to prevent the imminent danger of prior seizure of this strategic location by the French. "This must be the key to all the fortifications in the land as far as Tierra Nova [Newfoundland, Labrador, etc.]. This power . . . it is absolutely necessary that your majesty should hold and be master of . . . and if this arm of the sea, as is certainly believed, does go to the South Sea, it. is near China."[88] Menéndez promised to establish a fort with 200 men there.

Menéndez heard of a westward passage from several sources.[89] A garbled version reached the French court from Raimonde de Beccarie de Pavie, baron de Fourquevaux, the French ambassador to Philip II, who reported that Juan Pardo's expedition of 1566 to and beyond the Appalachians went "on a brig up the St. Helena river for about a hundred leagues. . . . The opinion of these explorers is that the river leads to Canada and that there is a passage through to the South Sea and to China. I hear that Pedro Menéndez too maintains the truth of this."[90] This story, spread in France through various channels, may be one of the sources for John White's general manuscript map of the Southeast, which features a strait from Port Royal connecting with a large, unnamed bay or sea to the west.

If Menéndez's general conception of North America was based upon false reports, his knowledge of the southeast coast, its harbors, and its strategic places for fortifications grew rapidly. The southeast littoral had been frequented and known to European sailors for many years; Velasco, the viceroy of Mexico, urging Tristán de Luna in 1560 to establish settlements, wrote: "The French come quite near to Santa Elena nearly every year to buy from the Indians."[91] Repeatedly he advised the use of John, an Englishman then married to a French woman, who as a cabin boy in 1541 had visited "the land of Florida in 37° [where] they found a very good bay [the Chesapeake]" and traded "a thousand martens in exchange for knives, fishhooks, and shirts."[92] "He has been there several times to trade . . . in the country northward of Santa Elena."[93]

28

By the summer of 1566 Menéndez had established himself at St. Augustine and extended his fortifications as far as Santa Elena. He then decided that the time was ripe to place an outpost on the Baya de Santa Maria. The Indian Don Luis had assured Menéndez and some Dominican friars of the certainty of mass conversion to Christianity of all Indians under his three brothers, who were chiefs of a territory, he claimed, that stretched from 36 to 39 degrees north latitude. On August 14, 1566, Menéndez dispatched the vessel *La Trinidad,* which carried two Dominican friars; Don Luis; an initial supporting force of fifteen soldiers; and a trusted Azorean pilot, Balthasar Fernándes of Tenerife, who may have known the coast. The ship made landfall "at the mouth of a river at 37° 30'" (Chesapeake Bay?) but was driven out to sea by a hurricane. Ten days later the vessel regained the coast but found itself at a river at 36 degrees north (an inlet into Pamlico or Albemarle Sound?), where its crew landed and took possession of the land in the name of Philip II of Spain.[94] During a seven-league exploration up the "River" (Albemarle Sound?), Don Luis failed to recognize any landmarks in the terrain; nor did anyone see the natives, possibly because a gathering storm interfered with fishing. *La Trinidad* was again blown out to sea; it returned again to the river at 37 degrees 30 minutes, only to be driven out to sea a third time. On a vote, the ship's crew decided to abandon the mission. They sailed for Spain, reaching Cadiz on October 23 and reporting to the Casa de Contratación in Seville a week later.[95]

The plan of Menéndez to make the Indians Christianized allies against other European intrusion and to explore the passageway through Rio Salado was thwarted. In 1570 a Jesuit mission was established with Menéndez's support, probably on the York River in present-day Virginia. But Don Luis soon abandoned the Jesuits, reverted to native customs such as polygamy, and, infuriated by the priests' vigorous remonstrances, set upon them with his followers and murdered them. Only a young Spanish acolyte was protected and adopted by a chief, an uncle of Don Luis. The crew of a supply ship, dispatched from Havana by King Philip's order in the spring of 1571, found no trace of the Jesuit fathers but learned from captives taken that something tragic had happened. In 1572 Menéndez, returning to Spain from his last visit to Florida, entered Chesapeake Bay; succeeded in rescuing the acolyte; learned from the boy the details of Don Luis's treachery; and taught the Indians of the region a characteristic punitive lesson by capturing, then baptizing, then hanging a number of them. He then sailed with his fleet for Spain, where he died two years later without having explored the upper Chesapeake for the passageway to the South Sea. In 1588 Pedro Menéndez Marqués, Avilés's successor, sent to the Chesapeake Captain Vicente González, the pilot for several previous voyages there. González reached the Susquehanna River and reported finding "a river which Captain Vicente Gonzalez believes for certain, goes in to the South Sea."[96]

Two new names are used in reference to the region surrounding the location of the Jesuit mission: Ajacán and Bahia de Madre de Diós. Ajacán (Jacán, Axacám, etc.) apparently was the name given by the Indian Don Luis de Velasco to the general territory from which he came; it was generally accepted as such by the Spanish writers. Bahia de Madre de Diós (Bay of the Mother of God) was applied more specifically to the Chesapeake. The Jesuit mission was at the Bahia de Madre de Diós de Ajacán, although the earliest documents frequently refer to the comings and goings of the Jesuits to Baya de Santa Maria. It is possible that Santa Maria was felt to be a more general and confusing name because it included the Carolina sounds as well; many maps showed the row of islands across an open, shallow bay at about 36 degrees north latitude instead of a narrow 3- or 4-league-wide mouth at 37 or 37½ degrees. The new names never became generally used on maps, at least those that have survived. Clifford M. Lewis and Albert J. Loomie, authorities on the Spanish Jesuit mission in the New World, note only one map with the name Jacán (B: de Jacam): a Portuguese Sanchez 1618 chart of the Atlantic coasts,[97] and not until almost a century later "Barra de Madre de Diós," at Chesapeake Bay on a Dutch chart: Jansson's Mar de Nort, ca. 1655, with a few later Dutch examples.[98] These rare variants could have been and probably were picked up from some printed report of the Menéndez-Jesuit mission period.

What cartographic changes, one may ask, resulted in European, especially Spanish, maps from Spanish exploratory voyages, reports, and attempted settlements along the North American coast, or even from other European trading and privateering activities? One can find surprisingly little change, and even less improvement, in nomenclature or topography. For charts deriving from Spanish sources, this is probably not strange; both Menéndez de Avilés and the Spanish government wanted as little as possible known about harbors or plans for their fortifications. If the policy of Spain and Portugal was to prevent charts from leaving the hands of trusted pilots or officials, they were admirably successful. Few significant changes were made in the maps and atlases produced by Iberian map makers in the latter half of the sixteenth century.[99] If the hydrographic department of the Casa de Contratación kept its padrón general updated, or produced charts showing improved knowledge of the coastal eastern seaboard below Canada, the charts have not survived.[100]

A fortunate exception is an undated, unsigned map of the Caribbean and eastern North American seaboard (10 to 42 degrees north latitude) found among the papers of the great chief cosmographer Alonso de Santa Cruz at the time of his death, November 9, 1567.[101] One of the famous group of hydrographers that established the preeminence of the Casa de Contratación in that century, Santa Cruz had returned in 1530 after four years of exploring South America with Sebastian Cabot; was immediately appointed cosmographer for the training and examination of pilots in the Casa; aided Chaves in the revision of the padrón general of 1536; and remained active while Sebastian Cabot, Diego Gutiérrez, and others were pilots major.

At the time of his death, Santa Cruz was still the active cosmographer major and senior lecturer in navigation.

The significance of the map is that it has "r: de s: agostino" at 29 degrees 55 minutes north latitude and "r: de s: matte" at 30 degrees 30 minutes and therefore must be dated after Menéndez landed and named St. Augustine and St. Matthew (St. John's River) in 1565; also significant is the fact that the map includes the rivers "St. Age," "St. Andres," "guales," "tatatecoro," etc., which Pedro Menéndez de Avilés first visited while on his way to establish a fort (St. Felipe) at Santa Elena in the spring of 1566.[102] The map could not have been drawn before 1566, therefore; and, given the time that must have elapsed before reports reached Seville and were analyzed by Santa Cruz or in the Casa, probably not before 1567. If the map was drawn by the ill or dying Santa Cruz, as is probable inasmuch as it was found with his personal possessions, that may explain a few obvious errors such as "Trafalcar" for Trafalgar (Cape Lookout) and "alsa piglago" for "arcipiellago," as Santa Cruz had himself written in 1545 for the purpose of describing more correctly the islands in Casco Bay, Maine.[103]

Probably the most noticeable change from the maps produced by Santa Cruz since the 1540s is "B^a de n^{ra} Senora," the first change in the name of the bay Baya de Santa Maria since it first appeared, and the omission of all islands at the mouth of the bay or within it. The three names, including Madre de Diós, are, of course, synonymous (St. Mary, Our Lady, and Mother of God); the names are frequently used, but their sequential application to the same general region is rare indeed in Spanish maps. Coupled with Santa Cruz's omission of islands, it may be another indication that the Spaniards were restricting this designated area to the Chesapeake rather than including it with the Outer Banks and the sound islands. The shift from the early conception of an open bay to Chesapeake as an almost landlocked harbor was being caused by the explanation of the geography by Indians such as Don Luis and by those who had visited it.

Elsewhere the coastal contour from Florida to Penobscot Bay remains much as it was in the padrón general type of 1536 in the productions of Spanish and Portuguese map makers through the remainder of the century and even beyond.[104] It should be noted that Spanish cartographers of the Casa did not succumb to the theories of water routes across to the South Pacific either at the St. Lawrence or at St. Mary's Bay. The North American continent remained on both Spanish and Portuguese maps unimaginatively wide and solid from Cape Breton and Cape Trafalgar to California.

VI. The Dee and Fernándes Maps of 1580

What maps are most likely to have been taken on the Ralegh voyages? Any ship crossing the Atlantic to a comparatively new destination is likely to have had a map on board. Each ship in a convoy or expedition required a chart in order to enable it to reach a common destination in the event of separation from other vessels, delay for repair, or engagement by an enemy. Governor Ralph Lane referred to a place on a Spanish map after the arrival of the Grenville-Lane expedition in "Virginia" in 1585.[105] There is no positive identification in the records of which maps were taken by this expedition. There are, however, two large charts of the North Atlantic coasts that have a tantalizingly close relationship to the planners of the voyages. At one time both were in the possession of Dr. John Dee, both eventually reached the queen's map collection, and both are presently held by the British Library. One of these charts is a copy of a map owned by Simon Fernándes, the pilot major of the Roanoke voyages, who may also have had a hand in designing the other chart.[106] The second map,[107] signed and dated 1580 by Dr. Dee, was prepared while he was actively participating in the North American colonizing plans of Sir Humphrey Gilbert, Walter Ralegh, and others.

Dr. Dee, a learned mathematician and correspondent of European savants, suffered because his wide-ranging curiosity led him into the study of astrology, alchemy, and occult mysteries. Many years earlier he had entered the service of Queen Elizabeth I, at her command made an astrological calculation of an appropriate day for her coronation, was consulted on her illnesses, and gave her lessons on the mystical interpretation of his writings. He was a vigorous investigator of potential routes, eastward and westward, to Cathay, where, he believed, Chinese wise men possessed alchemic powers such as the ability to change base metals into gold. In 1580 Dee made a chart of a northeast passage to Cathay and gave it, along with a set of instructions for navigating to Cathay, to the English explorers Charles Jackman and Arthur Pet, who were attempting to discover such a route on behalf of the Muscovy Company. Dee's later chart of the Northern Hemisphere, made for Sir Humphrey Gilbert in 1582, is in effect a compilation of the various contemporary theories and attempts to find a westward passage; it has previously been mentioned.

Dee was regarded by the common people as a magician and had became increasingly preoccupied with the occult. At one point a mob broke into his house at Mortlake on the Thames near London while he was abroad in Poland and destroyed his furniture, library, and laboratory equipment. He died in poverty years later. Although Dee served for some thirty years as a kind of court scientist, his occult studies, horoscopes, and laboratory experiments were less important but related aspects of his work. Modern studies have demonstrated his erudition and many contributions, particularly in the field of philosophical geography.[108]

Dr. Dee's 1580 chart of the North Atlantic (plate 18) is a composite that apparently incorporated the best maps available to him or his designer. The southeastern coast, basically of the Spanish padrón-general type, is similar to the maps of Diogo Homem, whom Dee may well have known; Dee's chart, however, includes place-names not found on the Homem maps. Dee's chart shows a promontory at present-day Cape Canaveral that is very similar to and appears to have been taken from Homem's 1558 map;[109] the atlas in which it is found was made for Queen Mary in London and presumably was accessible to Dee in Queen Elizabeth's map collection. Dee's chart includes a legend off the Florida coast that is the earliest reference to the Gulf Stream noted on any extant map: "Canalis Bahamae versus Septentr. semper fluit" (The Bahama channel flows ever northward). Professor W. F. Ganong, who notes that Gerhardus Mercator's 1569 map of the world is very similar to Dee's chart in regard to the St. Lawrence River but differs from it in regard to New England, points out that "Dee's unknown compiler was learned in earlier cartography and of rarely good judgment."[110] On Dee's chart the opening for the Bay of St. Mary is low—35 degrees 20 minutes to 35 degrees 35 minutes north latitude—with several islands, including one at the opening; to privateers the bay thus looked attractive as a refuge from pursuit. Of this chart, R. A. Skelton wrote that Dee "presented a copy to the Queen and perhaps another to Ralegh."[111]

On October 3, 1580, Dr. Dee presented this chart of the North Atlantic to the queen. In November the pilot Simon Fernándes visited Dee at Mortlake, bringing with him a large chart that Dee turned over to an assistant to copy. Without title or date, this copy bears an inscription in the upper left center: "This cownterfet of Mr. Fernando Simon his Sea carte which he lent unto my master at Mortlake, A° 1580 November 20. The same Fernando Simon is a Portugale, and borne in Tercera being one of the Iles called Azores." This "cownterfet" (see plate 19), usually referred to as the Fernándes-Dee 1580 map, shows the coasts bordering both sides of the Atlantic Ocean and the west coast of the Americas from 68 degrees north latitude (i.e., Ireland) to 36 degrees south latitude (i.e., the Cape of Good Hope), with their bays, rivers, islands, and shoals carefully delineated but with the omission of all toponymy except that of the Gulf Coast and eastern North America. The original must have included nomenclature for all the given coasts.[112]

The "sea carte" is of a somewhat earlier padrón-general tradition than the Dee 1580 chart but was characteristic of the Portuguese maps of the third quarter of the sixteenth century, as evidenced by the wide entrance of St. Mary's Bay at 35 degrees 20 minutes and the shape of Cape Cod. The bay features two rows of three islands each dotted across its width. There is no river (Rio Espiritu Santo) entering the bay from the southwest, as on most Spanish maps. To the north of the bay is a large, wide extension with the name R. Saluador. This is apparently a scribal error for R. Salado; "Salt River" probably represents the long northern reach of Chesapeake Bay. The coast shown on the chart between C. Trafalgar (Cape Lookout) and the rows of islands (the Outer Banks), with one indentation, is presumably the islands, with the narrow Core Sound behind them, between the cape and Ocracoke Inlet. Captain Arthur Barlowe, one remembers, wrote that his expedition "sailed along the same, a hundred and twentie English miles, before we could finde any entrance, or river issuing into the Sea."[113]

The usual Spanish nomenclature of the chart is interrupted by a few Portuguese spellings (tierra lhana, low land) and anglicisms (R.S. John for R.S. Iuan or Ioan), which were probably made by a Portuguese map maker and by Dr. Dee's assistant.[114] The only new name is C. de S. Tomas, just above S. Elena, probably a transcribing error for the usual C. de S. Roman.

Simon Fernándes,[115] chief pilot for the first, second, and fourth Roanoke voyages, must have had more knowledge of the southeastern coast than is indicated by the chart he lent to Dr. Dee, even if only from reports of other pilots or sailors or through some cursory unrecorded voyage of his own. Fernándes certainly played a significant part in the location and establishment (and, arguably, in the eventual failure) of the Roanoke colony. Born about 1538 in Terceira, he was trained as a pilot in Portugal and then, as had many other Portuguese pilots before him, shifted his services to Spain.[116] While in the service of Spain Fernándes acquired an exceptional knowledge of navigation in the West Indies; this knowledge was later used to England's advantage and Spain's detriment. It has been suggested that he acquired pertinent information from Domingo Fernándes, Pedro Menéndes de Avilés's trusted pilot on the 1566 exploratory voyage that landed on the Carolina banks, or from others aboard the vessel, such as the Azorean pilot Balthasar Fernándes of Tenerife.[117]

By the early 1570s Simon Fernándes had turned to piracy; his activities were often associated with English freebooters, especially the notorious John Callice. Flagrantly avoiding customs in the sale of his booty in England, Fernándes was finally arrested and sent as a prisoner to London in 1577 for trial before the judges of the High Court of Admiralty. He was dismissed and freed, although the Portuguese ambassador had produced evidence "enough to hang him," including the charge that, in seizing a Portuguese caravel, he had killed seven sailors with his own hands. Fernándes's knowledge of Spanish trade routes and shipping, in the opinion of certain high officials, including the secretary of state, Sir Francis Walsingham, was apparently too valuable to warrant the loss of his active navigational services in behalf of England. Fernándes entered Walsingham's service and became known as "Mr. Secretary Walsingham's man" while serving as pilot or master for various expeditions that Walsingham approved or supported.

In 1578 Fernándes was chosen by Sir Humphrey Gilbert as pilot for a ship captained by Gilbert's younger half brother, Walter Ralegh. The voyage subsequently undertaken by this ship did not involve a crossing of the Atlantic, but it did instill in Ralegh a high regard for Fernándes's ability and knowledge. Shortly after the return of the 1578 voyage Gilbert dispatched Fernándes in the spring of 1580 across the Atlantic to explore the North American coast for suitable colonizing sites, an original purpose of the 1578 expedition. But this time Fernándes went as captain in a small 8-ton ship with a crew of ten. He returned in three months; the unusually short duration of the expedition and Gilbert's then current interest suggest that it was a trip to New England. Such a rapid voyage would be extraordinary at any time and must have redounded to Fernándes's reputation. Don Bernardino de Mendoza, the Spanish ambassador in London, wrote to Philip II in 1583: "Simon Fernando . . . is considered one of the best pilots in the country."[118]

There is, however, a darker side to Fernándes's reputation. Mendoza had previously written to Philip on June 3, 1578, that "one Simon Fernandez [is] a thorough-paced scoundrel who has been giving them [Gilbert] much information about that coast [the West Indies] which he knows very well. As I am told, he has done no little damage to the King of Portugal, by reason of the losses suffered in this kingdom by his subjects."[119] Fernándes was also anathema to the Spanish as "a heretic"—a Protestant—and at least one scholar has speculated that he was a "converso," i.e., a Jew who converted to Christianity.

Fernándes's English associates likewise found in him unsavory qualities. Apparently he was unusually blasphemous, vulgar, expansively conceited, and self-serving. He had few moral restraints. Piracy, especially if directed against the Spanish, was an irresistible attraction that dominated his plans and conversations. In 1582-1583 he was copilot of the galleon *Leicester,* flagship of a fleet planned for trade in the Moluccas, with Edward Fenton captain general of the expedition. Richard Madox, chaplain of the *Leicester,* kept a diary in cypher that portrayed the character of the officers of the vessel, often mordantly. Madox regarded Fernándes, who persuaded Fenton to abandon the East India trading mission in favor of piracy, as the evil corrupter of the voyage.[120]

Beyond the details on the chart, with its frequently misleading geographic features, that Fernándes lent to Dr. Dee to copy, what did Fernándes know of the Carolina-Virginia coast, either from personal exploration or through information gathered from others? His decisions as chief pilot for the Roanoke voyages, one gathers from the narratives, were usually followed, although he was frequently regarded as opinionated, obstinate, and faulty in judgment. A review of the Roanoke voyages for which Fernándes served as pilot and/or master of a ship (in 1584, 1585, and 1587) will be helpful in examining his part in them as well as giving an account of the sequence of discoveries made by the colonists.

During the first voyage of Philip Amadas and Arthur Barlowe, sent by Walter Ralegh in 1584, Fernándes was the pilot and master of the admiral (the flagship). Barlowe's account of the expedition's arrival on the coast of North America has previously been quoted; his enthusiastic and well-written narrative, evidently based on the ship's log, was almost certainly edited for propaganda purposes by Ralegh or by Richard Hakluyt, who published it in his *principall navigations voiages and discoueries of the English nation* (1589). After declaring that the expedition followed the coast for "a hundred and twentie English miles"[121] without finding an opening, Barlowe continued: "The first [entrance] that appeared unto us, we entred, though not without some difficultie, and cast anker about three harquebushot [about 800 to 1,000 feet] within the havens mouth, on the left hande of the same. . . ."[122] There the crewmen landed and took possession in the queen's name.

Which inlet these crewmen entered and precisely where along the coast it was have been the subjects of extensive and contradictory comments, inasmuch as the descriptions of the land they sighted and of the haven they found are vague and ambiguous.[123] A number of inlets are shown on the maps subsequently made by the colonists; but, in the words of Governor Ralph Lane, in a letter of August 12, 1585, written to Sir Francis Walsingham from Roanoke during the second voyage, a more careful search of the coast revealed "only in all iij [3] Entryes, and Portes," and these "not with grete shippinge at eny hande to bee delte with."[124] Lane then gave the most careful description of the three inlets that has been preserved:

Trynyty harboroughe ys one, and only of viij [8] foote Vppon the barre at hyghe water. . . . The other Ococon [Wococon] in ye Entry whereof all our Fleete strucke agrounde, and the Tyger [the flagship of the expedition] lyinge beatynge vppon ye shoalle for ye space of ij [2] houres by the dyalle, we were all in extreme hasarde of beying casteawaye . . . and soo with grete spoyelle of our provysyones . . . havynge abydden by iuste talle above, 89, stroekes agrounde. . . . The iij [third] Entry and the beste harboroughe of all ye reste, ys the Porte which is called Ferdy-Nando dyscoverdde by ye master and Pilotte maggiore of our Fleete your honors servante Symon FerdyNando.[125]

Every one of the major Carolina banks inlets shown in contemporary drawings, as well as other inlets that have since closed, has had its supporters as the 1584 entry. "Trynyty harboroughe" in Currituck Sound, a shallow, narrow body of water stretching north from Albemarle Sound behind the barrier islands, has had its proponents. The de Bry engraving of a lost John White drawing entitled "The arriual of the Englishemen in Virginia" (plate 26), which will be discussed more fully, shows a pinnace approaching Roanoke Island from the direction of Trinity harbor; Barlowe, describing the approach of the Englishmen, wrote: "Myself with seven more went twentie mile into the River they call Occam [Albemarle Sound]; and the evening following we came to an Island, which they call Roanoak,

36

distant from the harbour by which we entred, seven leagues: and at the North ende thereof, was a village of nine houses."[126] Although the distance between Trinity and Roanoke makes this area the possible landing place, neither Barlowe's statement that from the entry island "Wee behelde the Sea on both sides to the North, and to the South, finding no ende any of both waies" nor his assertion that the island on which they landed was "large stretching itself to the Weste" is applicable to a Currituck Sound entrance.[127]

David Stick, who has studied and written about the Outer Banks for decades, wrote in 1958: "They [the Englishmen] found an entrance through the Banks above Cape Hatteras, probably at present-day Jeanquite Creek north of Kitty Hawk."[128] This suggestion includes a wide range; but nearness to Roanoke Island and impediments to any long view north or south would seem to exclude any inlet north of Kitty Hawk. Theodore Caplow of the University of Virginia, in a paper read at the 1981 meeting of the Society for the History of Discoveries, came to the conclusion that Barlowe's entrance, shown on both of John White's manuscript maps, was at the inlet immediately north of modern Cape Hatteras. His analysis is carefully reasoned: Crotoan Island, immediately south of the inlet, extends to the west, and the expanse of water extends north and south on both sides, as can be seen by looking at the maps. The distance to Roanoke Island, however, especially to the northern end, is incompatible with Barlowe's description, and neither Lane nor any other contemporary mentions this inlet as navigable.

Historians David Beers Quinn and Samuel Eliot Morison place Barlowe's landing at an inlet found by and named after Simon Fernándes. This certainly became the main entry used by the colonists in 1585 and afterward to reach Roanoke Island, to which it provided the nearest and deepest access. Both Quinn and Morison suggest that Port Ferdinando may have been found during the 1584 voyage.[129] Quinn notes that Philip Amadas and Arthur Barlowe may first have landed in the vicinity of Nags Head or farther north, been discouraged in an attempted entry of Trinity harbor by the shoals at its bar and in Currituck Sound, and then found and entered Port Ferdinando. The location of Port Ferdinando in 1585 is now generally low and swampy, with no evidence of the former presence of an inlet. In 1846 a new entrance, Oregon Inlet, opened a few miles southward; it remains the principal outlet for navigation from the northern sounds. No inlet fills the description of Barlowe's landing place satisfactorily, but Port Ferdinando ("Hatorask" on the 1591 White-de Bry engraving) can be accepted with no problems arising in interpretation of other incidents in Barlowe's narrative.

There is, however, another surviving description of the landing of the Englishmen with a different account of what then transpired. Arthur Barlowe stated that after landing on July 13 and exploring the region, including Roanoke Island, he and his men made friends with the Indians and persuaded two of them—Manteo and Wanchese—to return with them to England. After about a month Barlowe and his men "resolved to leave the Countrey and to apply ourselves to return to England, which we did accordingly"—by the middle of September. But in the Archives of the Indies is an undated deposition made in Spain some years later by a corporal on the 1584 voyage, Richard Butler, who stated that the Englishmen landed "at a place called Ococa [Wococon]," disembarked twenty leagues farther on at a place "known to the English as Puerto Fernando but to the savages as Ataurras [Hatorask]," and then went twelve leagues north to a port with a 9-foot bar that the natives called "Chaco Peos" [Chesapeake?]. They then went to Bermuda and the Azores in an unsuccessful search for prizes and thereafter returned to England following a nine-month voyage. If Butler did not confuse events of different voyages or was not inaccurate as a result of the passage of many years, his account substantiates the landing at Port Ferdinando but also adds a prior landing at Wococon and an entry into Chesapeake Bay, about which he commented in a somewhat confused manner on intertribal hostilities.[130]

During the second voyage in 1585, which included seven vessels carrying 500 people, of which 108 were colonists, Simon Fernándes was the pilot major and master of the flagship, the queen's *Tiger.* The general of the fleet, Sir Richard Grenville, developed a strong antipathy to Fernándes, but the governor of the colony, Ralph Lane, defended him to Walsingham against the attacks of Grenville and others. Lane wrote that Fernándes "trewly hathe carryed himselfe with greate skylle, and grete gouernement all thys voyeage, notwithstanding thys grete crosse to vs all."[131] The immediate basis for condemnation of Fernándes was the grounding of the *Tiger* during an attempt to steer the flagship through the difficult passage of Wococon Inlet upon the arrival of the expedition. The journal of the *Tiger* attributes this misfortune, which caused extensive spoilage of precious food and goods before the vessel could be floated, beached, and repaired, to "the vnskilfulnesse of the Master," although possible extenuating circumstances are not recorded.[132]

The location of Wococon (or Ococon) Inlet explains Fernándes's directing Grenville's fleet there and the attempted entry. On the Fernándes-Dee 1580 chart (plate 19) the southernmost entrance to the Baya de Santa Maria is 35 degrees 20 minutes north latitude; on the Dee 1580 chart (plate 18) it is 35 degrees 30 minutes. Wococon is modern Ocracoke Inlet (35 degrees 4 minutes); in 1585 Ocracoke was probably nearer 35 degrees 5 minutes. That the leaders of the colony thought they were at the Baya de Santa Maria is confirmed by the sketch map sent back to England shortly after their arrival. On the map (plate 22) the unnamed inlet bears the legend "the port of Saynt Maris wher we arivid first." In addition, Lane's letter of August 12 to Sir Francis Walsingham includes the following reference to the inlet: "Thys Porte in ye Carte ys by ye Spanyardes called St Marryes baye."[133] Baya de Santa Maria on the two 1580 maps is half a degree north of Cabo de Tralfagar (Cape Lookout); the shoreline distance from Cape Lookout to Ocracoke Inlet is about 37 miles.[134] Whatever unrecorded maps or other information Fernándes may have had, it was probably the two charts in Dee's possession (in 1580) that guided the plans and decisions for the expedition that had been made in England prior to the departure. The maps and drawings made by John White during 1585 and afterward will be discussed later.

Governor Ralph Lane was active in dispatching parties to "search out" the surrounding country and led several himself. During one trip he heard about a bay to the north (Chesapeake Bay) from an Indian chief on Chowan River who described a land route to it; the bay was said to contain an island with a deepwater approach.[135] Lane sent an exploring expedition (whether by land or by small boat is not clear), which apparently wintered (1585-1586) with the "Chesepians," a group of Indians situated about fifteen miles west of Cape Henry, apparently on Lynnhaven Bay. There is no evidence that Lane ever identified the bay to the north as the Spaniards' Baya de Santa Maria; he thought the Roanoke River was the Rio Salado, with its headwaters near the South Sea, and rowed for three days from its mouth to a point "160 miles from home" before turning back.[136]

Lane was so favorably impressed with the reports of the bay to the north and the fertile soil there that he recommended moving there from Roanoke Island after an intended visit to find a suitable site in the summer of 1586. Lane later became discouraged with prospects for the future success of the colony or its relocation, but his and White's description made Hakluyt and others vigorous supporters of such a move.[137] Ralegh's written orders in 1587 were to plant the colony in the Chesapeake area.

The colony under Lane came to a sudden and unexpected end with the visit in June, 1586, of Sir Francis Drake during his return trip from his "Famous West Indian voyadge." Drake offered the colony ships and additional settlers, but he lacked sufficient additional supplies to give them. David Beers Quinn has shown that Drake's proffered settlers were probably Turks, Moors, blacks, and others, liberated prisoners and galley slaves collected during Drake's attacks on Spanish Caribbean ports.[138] Grenville, whose arrival with reinforcements of colonists and supplies had been expected by early spring, had not come. A severe storm destroyed Drake's proffered ships, and the discouraged settlers voted to return to England with Drake. Grenville's fleet arrived soon afterward; Grenville left a small body of men to maintain continuity of settlement and likewise returned to England with a booty from privateering. No significant cartographic records appear to have survived from these first two Ralegh expeditions.

During the principal and final attempt at settlement in 1587, John White served as governor and Simon Fernándes as pilot and master of the flagship *Lion.* Under Governor White Fernándes also served as one of the twelve assistants of the chartered "Cittie of Ralegh," having been granted arms with the status of gentleman. The instructions for the new colony were to visit Roanoke Island briefly, interview the fifteen men left there by Grenville, and establish a small base for the support of privateering. Governor White was then to "returne again to the fleete, and passe along the coast, to the Baye of Chesepiok, where we intended to make our seate and forte, according to the charge given vs among other directions in writing, vnder the hande of Sir Walter Ralegh."[139]

White, with forty men, landed at "Hatoraske" (White's later name for Port Ferdinando) expecting to rendezvous with the men left by Grenville in 1586 and then reboard for the Chesapeake. With the exception of the bones of one man, no sign of Grenville's holding force was seen. From Indians on Croatoan Island White learned that Grenville's small force had left in their pinnace for England. They never arrived. The settlement at the north end of Roanoke Island was deserted. At this point Fernándes, through a spokesman, forbade the sailors to bring any planters back to the ship: "he would land all the planters in no other place . . . saying that the summer was farre spent."[140] The season for privateering—Fernándes's obsession and probably that of his crew—was nearly over. The hurricane season began in August, and the Spanish fleet would soon be starting on its way across the Atlantic.

Yet, Fernándes's refusal to follow Ralegh's orders to take the colony to the Chesapeake is difficult to understand. He did not depart until August 25, more than a month after the arrival at Port Ferdinando. His actions during the voyage from England, concerning which only White's critical account is extant, seem erratic. Fernándes abandoned an accompanying flyboat under Captain Edward Spicer that was equipped with provisions and colonists and was angered when the vessel reached Port Ferdinando instead of being lost. He refused to remain on a West Indian island long enough for colonists to gather plants and salt and almost wrecked the admiral on a barrier [Bogue] island west of Cape Lookout, which he claimed was Croatoan Island.[141] Probably the chief cause of friction and frustration that developed between White and Fernándes was the latter's desire to miss no opportunity to find and take a prize. White wanted no hazardous encounters to endanger the women and children aboard or to imperil the planting of the colony. The refusal of Fernándes to take the colonists to the Chesapeake may have resulted in part from his malice toward White, but he was undoubtedly anxious to leave the colonists and get on with his real interest. Taking them to a new site fraught with navigational dangers, finding a suitable location for them, and seeing to it that they were safely established involved unpredictable delays.

Fernándes left aboard the *Lion* to engage in privateering. White, at the written entreaty of the colonists, left the same day, August 25, with Captain Edward Spicer in the flyboat to obtain needed supplies from England. Fernándes reached Portsmouth in late October with no prizes taken; the flyboat reached Ireland in November after a protracted trip, with most of the crew sick or dying from injuries or starvation. White set out again in April, 1588, with provisions for the colony in two ships, but the ships were intercepted by a French vessel that stripped them of all but anchor and sails. The crippled ships reached England again with difficulty; the

40

impending attack of the Spanish Armada prevented additional voyages in 1588. White succeeded in reaching Roanoke in 1590, only to find the settlement deserted. The subscribers to the colony were unwilling or unable to undertake additional voyages or support, and White and the colonists soon faded into history.[142]

In light of Simon Fernándes's checkered relationship with the leaders of the English colonizing venture in North America (particularly Sir Walter Ralegh, who probably condemned him for not following his orders), the chief pilot's overall record is difficult to assess. Glimpses of Fernándes's continued vendettas against Spanish shipping crop up. He was boatswain under Captain Martin Frobisher aboard the *Triumph,* Elizabeth's largest galleon, in the running battle with the Armada, which culminated in the attack upon the *San Martin,* the Spanish flagship, at Gravelines. In 1590 he was master of the great 300-ton *Foresight* under Captain William Winters in Frobisher's and Hawkins's thirteen-ship fleet that attacked Spanish commerce. Fernándes probably ended his career attempting to take one prize too many.

During the Roanoke enterprise the effect of Fernándes's presence was both positive and negative. As chief pilot, he exhibited extensive knowledge of trans-atlantic navigation, the trade winds, shipping routes, and complex West Indian geography; this knowledge was a major factor in the successful crossings made. During the 1584 voyage he took Amadas and Barlowe to the latitude (35 degrees 30 minutes north) and location on his map—known by reason of a copy made by Dr. Dee's assistant—at which there existed a large open bay, the Santa Maria, with islands. He found Port Ferdinando, near present Oregon Inlet, where smaller boats could cross the inlet bar to Pamlico Sound.

On the other hand, Fernándes's specific previous or personal knowledge of the seacoast of southeastern North America, its harbors, bays, and headlands, appears negligible. Any interpretation of documents implying earlier exploratory voyages by Fernándes is open to question. Moreover, Fernándes failed to utilize any knowledge of contemporary Spanish cartographic information, although from at least 1566 onward, Spanish pilots referred to 37 degrees north as the latitude of Baya de Santa Maria.[143] After 1570 the Chesapeake was called Axacán or Bahia de Nostra Senora, possibly to distinguish it from inclusion with the Outer Banks coast. In addition, after 1570 at least, the general size and shape of the Chesapeake was known in St. Augustine because of numerous voyages sent there during the ensuing decade. Fernándes's chief interest and personal objective was to find a haven for privateering, which involved sorties on Spanish shipping returning by the Gulf Stream, quick disappearance from possible pursuit, the repair of ships, renewed supplies, and safe docking facilities. These the North Carolina coast could not provide, but the Chesapeake could. Ignorance of the Chesapeake or fear of Indian hostility, if he knew it existed, may explain the refusal of Fernándes to take the Roanoke colonists there.

41

It is possible that Fernándes may have heard of the massacre of the Jesuit mission of 1570 under Father Segura at Axacán by the Indian Don Luis and his tribe. Fernándes may have visited the bay himself after leaving Roanoke in 1584, if the deposition of Corporal Richard Butler is accurate: "from Puerto Fernando . . . they [Fernándes's party] moved twelve leagues to the north and found a port . . . Cacho Peos [Chesapeake?]."[144] The 1586 deposition of Edward, a castaway of the 1585 voyage, found by the Spanish on Jamaica, is ambiguous and confusing. Edward reported that the English came to a bay with some islands and two promontories. They landed on one, where hostile Indians ate thirty-eight Englishmen. The Portuguese (Fernándes) then took them to the other promontory, where the Indians were more friendly. (See note 130.) On what voyage and on what promontory (if not the 1585 voyage) the English were attacked by Indians is not clear.[145] No such cannibalistic occurrence is mentioned elsewhere in the voluminous records and reports relating to the 1585 voyage or any other Roanoke voyage. Had any such episode occurred, Governor Lane would certainly have learned of it from Philip Amadas, captain of the 1584 voyage and officer in 1585.

One of the most interesting contemporary comments that may indicate that Fernándes did explore the Chesapeake is found in William Strachey's *The Historie of Travell into Virginia Britanica*, written about 1612 following Strachey's return from the Jamestown colony. Strachey wrote: "There is extant an old plott, which his Lordship [Lord Delaware] hath shewed me, wherein by a Portugale our seat is layd out, and in the same 2. silver Mines pricked downe."[146] No Portuguese except Fernándes is known to have been with the English Roanoke voyages: if Fernándes did explore the lower Chesapeake, the "old plott" could be his cartographic report to Sir Walter Ralegh.

The reason for Fernándes's refusal to take the Roanoke colonists to the Chesapeake in 1587 is less likely to have been the danger of encountering hostile natives or doubts concerning safe harborage than reluctance to delay departure from Roanoke for privateering. It would seem out of character for Fernándes to have feared an encounter with Indians or to worry about placing the colonists in jeopardy.

VII. Islands, Sounds, Inlets: What the English Colonists Found

The maps and reports with which the planners of and participants in the Ralegh voyages were familiar or had with them had provided them with certain ideas about the coast they approached; these maps have been analyzed by this work in some detail. The Ralegh voyagers also made maps on the basis of their own explorations and observations; these maps are treated in chapter 8. The following discussion, a brief factual description of the North Carolina portion of the North American coast, will serve as a helpful introduction to this volume's final chapter, an examination of the cartography of Ralegh's Virginia.

Reality did not correspond to expectation when the English reached their destination at 35 to 36 degrees north latitude in 1584. There was no great open bay with scattered protecting islands, as shown on all the best Spanish and Portuguese charts. Instead, the English found an unbroken coast, along which, Arthur Barlowe reported, "wee sailed . . . a hundred and twentie English miles, before we could finde any entrance, or river, issuing into the Sea. The first that appeared unto us, we entred, though not without some difficultie. . . ."[147] The difficulty of entry, had Barlowe but known, presaged later disasters. The bark *Bonner*, offered the voyagers by Sir Francis Drake in June, 1586, could not enter, and the pinnaces that carried the colonists out to Drake's ships in the aftermath of their decision to return to England were repeatedly grounded, resulting in the loss of "most of all wee had, with all our Cardes, Bookes and writings. . . ."[148]

The navigational hazards and obstacles associated with all the inlets along the coast has been an unsurmounted problem for the commercial and industrial development of eastern North Carolina behind the sounds to this day. The shore that Barlowe saw was not, of course, the mainland as he first supposed,[149] but a long, narrow chain of barrier islands extending uninterrupted for some 180 miles north of Cabo de Trafalgar or Promontorium Tremendum (Cape Lookout). Behind these barrier islands lay several sounds, some of them narrow and island-studded, while two of them—Pamlico and Albemarle—extend westward from the coastal islands two dozen miles or more. These sounds were shallow and presented immediate hazards to seagoing vessels that might succeed in crossing their inlet bars.[150]

Except for a few small vessels such as pinnaces, the ships that brought the colonists were heavily armed for protection and privateering, and thus it was necessary to anchor them out in the open sea, unprotected, on one of the most dangerous and continually turbulent coasts in the world. Only Sable Island, rising in the Atlantic Ocean seventy miles south of Cape Breton, with its eleven-mile spit, vies with the Cape Hatteras area as the "Graveyard of the Atlantic"; both locales have been associated with hundreds of known shipwrecks, in addition to untold numbers of vessels lost without record, since the earliest years of the sixteenth century. Clifford M. Lewis and Albert J. Loomie enumerate wrecks in the vicinity of Cape Hatteras in 1528, 1545, 1551, 1553, 1559, 1561, and 1564.[151] A storm that struck the coast during Sir Francis Drake's visit in June, 1586, drove some of his larger ships out to sea, not to return. "Manie also of our small Pinnaces and boats were lost in this storm" is the terse report in Walter Bigges's *summarie and true discourse* of the expedition.[152] Even on good days there are perilous winds off Hatteras, which is surrounded by the treacherous Diamond Shoals. The *U.S. Coast Pilot* notes laconically: "Navigation in the area is extremely hazardous for all types of craft. Tropical cyclones in summer and extratropical storms in the winter plague the mariner. Hurricanes are often at full intensity when they pass Cape Hatteras."[153]

The Gulf Stream, moving northward along the Florida coast through the Bahama Channel at about sixty miles a day, is forced out to sea by the eastward projections of Cape Lookout and Cape Hatteras and by the deflecting force of the earth's rotation.[154] In addition, off Cape Hatteras and its Diamond Shoals the Gulf Stream meets the southward-moving Arctic or Labrador Current, which forces the warm flow farther eastward while the colder stream becomes a countercurrent that moves southward along the coast. The interaction of these two great countercurrents is a major cause of the constant ocean and air turbulence in the area.

The Labrador longshore current, with its strong accompanying winds, moves the sand, especially the dry dune sand on the Outer Banks, southward. This is why inlets regularly fill up with drifting sand on the north side of the banks and are scoured away on the south side, causing many inlets to fill up or move slowly southward down the coast. A wreck in an inlet often catches the drifting sand and closes the entry. The pressure of water may break through a bank island to form a new inlet, sometimes not far from the old.

The very existence of the Outer Banks is more difficult to explain; the banks resulted from slow geologic and climatic changes. Geologists generally believe that during the last Wisconsin glacial advance (about 40,000 years ago) the ocean level dropped markedly, resulting in the emergence of a shoreline far out on the present continental shelf. During subsequent warmer interglacial periods the ocean rose above its present level, causing submergence of portions of present eastern North Carolina.[155] This phenomenon can be seen in the so-called Pamlico and Talbert terraces, in which fossilized marine deposits and scarps, caused by erosive wave action, are present west of New Bern, Washington, and the Chowan River.[156] As geomorphologist A. K. Lobeck of Columbia University wrote, "The originally upraised coast of North Carolina and much of the Atlantic coastal plain was deeply embayed so that it is almost equally a shoreline of emergence and one of submergence.''[157]

The barrier islands of North Carolina were originally submarine bars or ridges (as are those of the Newfoundland Banks at present); such bars grow in height with the accretion of additional sand washed up by the longshore currents until they rise above the outer surface. The movement of such barrier banks is toward the mainland, and the westward movement of the Outer Banks varies from a few inches to several feet every year.[158] Save for this slow westward advance of the barrier islands and the general southward drifting of the sand between Cape Henry and Cape Hatteras, the general contour of the Outer Banks has changed little since the sixteenth century. An exception is the recession and disappearance of a cape about thirty miles north of Cape Hatteras. It is by far the most prominent cape on the John White maps and the later derivatives but is unnamed by White except as "the poynt that lyeth to the Southwardes of Kenricks mount," the place at which Captain Edward Spicer's boat lost its anchor and almost wrecked.[159] Avulsive action by sea and storm before the last quarter of the seventeenth century has left only a gentle outward curve, though the unnamed cape still extends farther east into the Atlantic than Cape Hatteras; beyond it lie the dangerous submerged Wimble Shoals.[160]

The inlets along the Outer Banks north of Cape Lookout vary in number from three to more than ten at any given time. On White's manuscript map of the coast, drawn in 1585, there are 9 inlets; the White-de Bry engraved map of 1590 shows 12; the current United States Coast and Geodetic Survey charts show 7. Only two of these inlets—Ocracoke and Oregon—are navigable at present. The low-lying Core Banks to the southwest include several inlets that are too shallow for passage. "The inlets themselves," David Stick has written, "are forever changing shape, size and depth, sometimes opening wider, at other times shoaling up and closing altogether, yet throughout it all maintaining a leisurely but steady movement down the Banks."[161]

The inlets are made and kept open not by tidal movement or ocean waves but by the constant and tremendous flow of water from the bays; this flow is fed by the tributaries and such rivers as the Chowan, Roanoke, Tar, and Neuse. The annual flow from these waterways into Pamlico Sound, according to a recent United States Geological Survey investigation, is nearly a trillion cubic feet from a drainage area of 31,900 square miles, in addition to the tidal movements.[162] When an inlet is shoaled and closed, the outflow of others is increased or another inlet breaks through at a place weakened or opened by a major storm. During a hurricane the counterclockwise movement of the tidal storm surge forces the bay waters westward up the estuaries and rivers, and the great ocean waves rush through the inlets or over low-lying Outer Banks islands. When the inevitable eastward-moving following wind comes, as the cyclonic storm moves northward, the mass of ocean and penned-up freshwater drainage tears through the inlets, enlarging them or making new ones.

In all, some thirty Outer Banks inlets have been recorded.[163] Some of these have opened and closed several times, and the same inlet may have had different names during its history: Port Ferdinando (White, 1585) became Hatorask (de Bry, 1590) and then Oregon, an inlet slightly to the south that was opened in 1846. Four inlets have been named "New." Many inlets have been too transitory to be named. Wococon (present Ocracoke) is the only Outer Banks inlet that has remained open continuously since 1585.[164]

Erosive action is not limited to the barrier islands; it also occurs within the sounds. Through this phenomenon, sound islands have changed in shape or disappeared. "Hariots Ile" ("Batts Grave" in the eighteenth century) in Albemarle Sound near the mouth of Yeopim River is shown on the White-de Bry map of 1590 and thereafter, but it gradually eroded away in the nineteenth century.[165] More significant is the erosion of the northern end of Roanoke Island, which has been washed by all the water flowing from Albemarle Sound and its tributaries into Pamlico Sound and to the outlets at Port Ferdinando (Oregon) and other inlets. Failure thus far to find the colonists' settlement and houses, which are believed to have been located near the verified site of the fort on the northern shore of Roanoke Island, may be linked to shoreline erosion. On the basis of verifiable erosion data for the last 120 years, geographers Robert Dolan and Kenton Bosserman have concluded that "The shoreline in the vicinity of the settlement has eroded as much as a quarter of a mile during the last 400 years." Dolan and Bosserman believe that "artifacts, if any, have been corroded, dispersed, and buried in the waters adjacent to the present shoreline."[166]

The problems that stood in the way of the successful planting of the Roanoke colony, it must be observed, did not stem entirely from geographic ignorance, crucial as the choice of location turned out to be.[167] Even if the colonists had reached a good harbor, such as Chesapeake Bay might have afforded, other existing or potential difficulties would probably have endangered or prevented a permanent settlement. The motivations of the backers of the Ralegh voyages were multiple; some of them were impractical if not mutually contradictory. The economic support for the undertaking was not widely enough based; the enterprise required more fiscal and official support than Ralegh and his associates could continue to provide. The inability of the colonists to find gold or mineral ore or a passageway to the South Sea was devastating to morale. "[A]fter gold and siluer was not so soone found, as it was by them looked for," Thomas Harriot wrote, [the colonists] "had little or no care of any other thing but to pamper their bellies. . . ."[168] Fertile soil for abundant crops was expected, but basic agricultural problems were not foreseen; at 35 degrees north latitude the Hakluyts and others anticipated a Mediterranean climate and produce. Simon Fernández and the sailors aboard the transporting vessels, as well

46

as the colony's backers in England, expected the establishment of a fortifiable haven complete with supplies and repair facilities to which privateers could return at the conclusion of prize-taking. The war with Spain, with the Spanish Armada as its apex, also had a part in the eventual failure of the Roanoke colony.

Some of these problems, of course, resulted from geographic ignorance of the region. Yet, two gifted men who accompanied the Ralegh colony produced maps and reports that for future English settlers left an extraordinary record of what they found. The cartographic productions of these men will be examined in the following chapter.

VIII. The Cartography of Ralegh's Virginia

Surveys of the coastal area between Cape Lookout and Chesapeake Bay made during the first English attempts at settlement and the maps that resulted are as illuminating as the earlier maps—based primarily on the voyages of the Licenciado Lucas Vasquez de Ayllón and Estéban Gómez more than a half century before—are inaccurate and misleading. The quality of the English maps resulted from the work and skill of two of the colonists: Thomas Harriot and John White. Harriot's many accomplishments included surveying, and White was an already experienced painter skilled at close and accurate observation. Their presence was not fortuitous: persons with such skills were especially recommended to Sir Walter Ralegh in planning colonization. No official documents giving the plans and orders that are known to have been made for the Roanoke voyages have been preserved. There are, however, several informed and detailed sets of instructions prepared between 1582 and 1586 for the effective organization of English overseas enterprises and colonization. Several of these instructions include lists of artisans and skilled professionals necessary for a successful venture.

One of these instructions, an unsigned set of notes, identified as having been made by Roger Williams, with the heading "For *Master* Rauleys Viage" and endorsed "Notes geven to *Master* Candishe," was probably written in 1584, inasmuch as Ralegh was knighted at Greenwich by the queen on January 5, 1585. Thomas Cavendish was high marshal and captain of the *Elizabeth* during the 1585 voyage.[169] "Then I would have [a] good geographer to make discription of the landes discovered, and with hym an exilent paynter," wrote the author, whose general emphasis on military personnel and fortification indicates that he was an experienced soldier.[170]

Another set of instructions for a voyage of reconnaissance to North America in 1582 or 1583 is associated with the plans of Sir Humphrey Gilbert for an expedition that was projected to depart sometime between June, 1582, and March, 1583, but which never sailed.[171] It includes detailed orders directing one Thomas Bavin to make surveys, to produce maps, and to "draw to life all strange birdes beastes fishes plantes hearbes [and] Trees." Of special pertinence are the cartographic methods and technical equipment elaborated in these instructions under numerous headings: "Let Bavin carry with him good store of . . . Instruments to draw cardes and plottes." This equipment included, among other items, paper royal, different colored inks, quills, and compasses. Bavin was instructed to have with him at all times an attendant to carry his instruments and other assistants with cross staffs and other types of surveying equipment. He was to employ rigidly in his field drafts a set of sixteen symbols for natural features such as woods, hills, rocks under water, rocks above water, etc.; the set of symbols is provided in the instructions and also on a separate chart.[172]

The importance of this set of instructions is that a copy of it may well have been known to and influenced the plans of Walter Ralegh, whose North American patent of 1584 followed the grant held by his half brother, Sir Humphrey Gilbert, whose colonial enterprises in North America were abruptly ended by his death in a storm at sea in August, 1583.

Another document should be mentioned here: "Inducements to the Liking of the Voyage intended toward Virginia in 40. and 42." degrees of latitude, written in 1585 by Mr. Richard Hakluyt, lawyer of the Middle Temple and, like his younger cousin of the same name, a strong advocate of overseas expansion. Hakluyt ended the propaganda tract, for such it was, "A skillful painter is also to be carried with you which the Spaniards used commonly in all their discoveries to bring the descriptions of all beasts, birds, fishes, trees, townes, etc."[173] The need for a scientist and a painter is recognized in these lists of skilled craftsmen and professionals; it is doubtful, however, that two such brilliant choices as Thomas Harriot and John White could have been expected.

White was probably present during the first voyage, and Harriot may possibly have been, although neither is mentioned in Arthur Barlowe's report as edited and published by Hakluyt.[174] White, in a letter to his friend the younger Hakluyt, written in 1593, declared that his unsuccessful voyage of 1590 was "my fift and last voiage to Virginia."[175] There are recorded only five voyages during which White could have been a passenger, including the exploration of 1584. During the second voyage, that of the first colony in 1585-1586, both White and Harriot played active and well-recorded parts. White produced most of his watercolor paintings of his observations, and Harriot gathered extensive notes, some of which were printed in his *briefe and true report of the new found land of Virginia* (1588), reprinted by Theodor de Bry in 1590 and by Richard Hakluyt in 1589 and 1600. White's third voyage was in 1587, when as governor of the second colony he returned to England for needed supplies after only a month's stay. In 1588 he embarked from England with two supply ships for a return to America, but the ships were captured by a French vessel, stripped, and returned to England with difficulty. White did not actually reach Virginia. The threat posed by the Spanish Armada resulted in the husbanding of all available English resources in ships and men. In 1588 the lure of rich prizes in privateering following the defeat of the Armada attracted English investors to shipping enterprises. The lack of funds forthcoming for a colony, with its diminished strategic value, prevented White from undertaking his "fift and last voiage" until 1590.

49

This brief survey reveals that White and Harriot had little opportunity for observation and work at Roanoke except during the existence of the first colony. The flagship *Tiger* arrived off Wococon (Ocracoke) Inlet on June 26, 1585; the colonists left with Sir Francis Drake's fleet on June 18, 1586. Harriot and White, each in his own way, produced a body of information about the New World remarkable in its scope and quality. It is generally agreed that the two men must have worked together, complementing each other's special abilities; but despite the voluminous records that have survived from that year's venture, little is stated, or even positively known, about the ways in which they cooperated—although Harriot's notes to White's drawings in de Bry give some intimations. The present knowledge of White's background and training for his tasks is very limited. Somewhat more is known of Harriot's early life and education, but large gaps remain there also.

Little is known of the early life or training of John White.[176] White was in London by 1584 and knew Walter Ralegh then or before. By 1578 he had apparently already made his drawings from life of Eskimos (presently in the British Museum with his portfolio of paintings of America).[177] The Eskimos shown in his drawings may have been those taken to London by Martin Frobisher in the course of one of his three voyages of 1576, 1577, and 1578 in search of a northwest passage; historian David Beers Quinn believes that White went with Frobisher to Baffin Island in 1577.[178] Two drawings in White's volume were of an Indian man and woman of Florida. These drawings are now identified as having been made by Jacques le Moyne de Morgues,[179] the French artist and map maker of René de Laudonnière's French Huguenot expedition of 1564, who escaped the massacre of the French colony on the St. John's River by Pedro Menéndez de Avilés in 1565, as previously discussed. Le Moyne had gone to England by the early 1580s during the bitter religious civil wars in France; he was given letters of denizen (a permit to establish residence) in London on May 12, 1581.[180] White may have introduced him to Ralegh at about that time. Le Moyne's drawings, as well as his knowledge of Florida, with its relative proximity to "Virginia," were of vital interest to the planners of the Roanoke enterprises, who mistakenly believed that overall conditions in the two places were similar. Ralegh commissioned le Moyne to produce paintings of Florida—presumably those Theodor de Bry saw while on a visit to England in 1587, acquired from le Moyne's widow in 1588, and engraved and published in 1591 along with le Moyne's written commentaries.[181] Le Moyne's influence on White's general map of southeastern North America will be discussed subsequently.

Of White's own training as a painter, nothing is known; the naturalistic style of his American drawings is markedly different from the English schools of the time and from his own Pictish or Oriental figures, which follow contemporary stylistic patterns. It is probable that, like le Moyne, White was neither a surveyor nor a cartographer; but if not, he was quick to acquire skills from Thomas Harriot and to adapt the information the two gathered during their journeys by boat and on land for the maps he designed and executed. These maps range from the plats of a fortified encampment and of entrenchments on Puerto Rico and of Indian villages in Virginia to the two maps of the North American coast.

Thomas Harriot's career both before and after the Roanoke voyages was primarily that of a mathematician, but his knowledge and activities encompassed a wider area in his study of and reports on the American scene. Born in Oxford in 1560, he entered St. Mary's Hall and received his B.A. in 1580.[182] Harriot's exceptional abilities were early recognized, for soon after finishing his studies he was recommended to Walter Ralegh and entered Ralegh's household to teach him "the mathematical sciences." Ralegh soon expanded Harriot's duties to "instruct [in navigation and astronomy] the sea-captains who later frequented his household"; by 1584 Harriot had written a course in navigation in his lost manuscript entitled "Arcticon."[183] Whether or not he went on the 1584 voyage with Philip Amadas is not known. In 1585, in accordance with special instructions, Harriot studied the Algonquian language, learning what he could of Indian culture and customs and, more importantly, reporting on America's resources in minerals, trees, plants, animals, birds, and fish. His coverage of these subjects in his *briefe and true report of the new found land of Virginia* and in the commentaries he wrote for Theodor de Bry's engravings of John White's paintings[184] closely follows the directions given to Thomas Bavin. Ralegh's instructions to Harriot and White must have been very similar.[185] In less than a year after arrival, Harriot had accumulated a wide and accurate knowledge of American conditions within the area surrounding the North Carolina sounds; this knowledge formed an excellent textbook for later English colonizing enterprises. Not all of his conclusions would be accepted at present: he recommended "drinking" tobacco smoke for its therapeutic value in curing numerous diseases.

Harriot's extensive "Chronicle" of the year in Virginia, which might have answered many questions now unanswered and added much to the present knowledge of ecology and conditions in pre-colonial America, has been lost. Harriot was an inveterate note taker; more than 10,000 sheets of his scientific notes are still in the British Museum, at Petworth House in Sussex, and elsewhere. Upon his return to England he remained in Ralegh's service until about 1598, when the ninth earl of Northumberland gave him lodgings in Syon House. During the long years of the imprisonment of Northumberland and Ralegh in the Tower of London, Harriot constantly visited his patrons, taking them books and aiding them in laboratory experiments.[186] He soon surpassed all other mathematicians in England, including Dr. John Dee and Anthony Wood, his reputed teacher at Oxford. He invented useful algebraic symbols; made improvements in optics, including observations of sunspots and moon spots with a 30-power telescope constructed in his laboratory; invented an improved backstaff for determining latitude at sea;

and made a list of recommendations for the 1607 Virginia (Jamestown) expedition. Harriot died July 2, 1621, in London.[187]

Sketch Map, ca. 1585 (plate 22)

The first cartographic record surviving from the Ralegh voyages is an anonymous undated sketch endorsed "A discription of the land of virginia," which encompasses Pamlico and Albemarle sounds and the surrounding terrain. The sketch includes fourteen legends that provide varied information concerning names of islands, Indian villages, and botanical and zoological specimens observed. Ocracoke Inlet, where the flagship *Tiger* went aground in 1585, is labeled "the port of saynt maris wher we arivid first," with the island of "wococan" to its north, identifying it as the entrance to the Bahia de Santa Maria of the Spanish maps. The English thought Pamlico Sound was the great bay; it was within a few miles or less in latitude of the entrance shown on the Fernández-Dee and other charts.

This rough sketch, presently in the British Public Record Office in London, was formerly and without justification (except for the endorsement "land of virginia") attributed to Captain John Smith during some nineteenth-century reshuffling of the State Papers. In an example of acute scholarly detection, historian David Beers Quinn demonstrated that the sketch could derive only from initial exploration after the 1585 landing and was probably sent back to England in the *Roebuck* or *Elizabeth* with Governor Ralph Lane's letter of September 8, 1585, to Sir Francis Walsingham, the queen's principal secretary and a major backer of the Roanoke voyages. The handwriting cannot be identified with the known script of White, Harriot, or Lane but may be a copy of an early draft of mapping done by White or Harriot.[188] It shows a generalized perception of the land and water contours, including Mattamuskeet Lake (Paquippe) and the surrounding swamp, discovered by Sir Richard Grenville's exploring party in July, 1585, and indicates directions to Indian villages reported but not visited. Simple as it is, the map reflects technical competence in its plain-line drawing of relative geographic distances and directions, its noting of natural features, and its use of symbols (as in open and palisaded villages).[189] The sketch has some relation to White's lost pictorial drawing of a slightly smaller area, preserved in the de Bry 1590 engraving entitled "The arriual of the Englishemen in Virginia" (plate 26), which will be discussed subsequently.

La Virgenia Pars [1585] (plate 23)

John White's general map of the southeast from the Cape of Florida to Chesapeake Bay, "La Virgenia Pars" is a remarkable rifacimento based on many maps examined previously in these pages.[190] It includes details from other maps and a new delineation of the Outer Banks region from the surveys of White and

Harriot. The ships, whales, dolphins, and flying fish in the Atlantic were ornamental additions perhaps seen by White in crossing; they recall the decorative style of the Flemish map engravers in England at the time,[191] but are more realistic—with flying fish chased by their pursuers—than the monsters found on many European charts.

The northern portion of the map, between Chesapeake Bay and Cape Lookout, is a smaller version of White's more detailed map of the Outer Banks region. It features fewer names of Indian villages, and Chesapeake Bay shows no northern shore, as it does on White's larger-scale map and the de Bry engraving of 1590. From Cape Lookout to Port Royal the coast, although incorrectly drawn on the same latitude, is more detailed than on any extant Spanish map and may suggest a cursory examination of the coast during the 1585 voyage. According to the *Tiger* journal, "we fell with the mayne of Florida" on June 20, "were in great danger of a Wracke" on June 23, made a landing and fished on the twenty-fourth, and "came to anker at Wococon" on the twenty-sixth.[192] Present Cape Lookout and Cape Fear are clearly delineated though unnamed, and farther down the coast are two unnamed bays, which are correctly located for Winyah and Charleston harbor.

At Port Royal (the Spanish Baya de Santa Elena) is a long, narrow river or strait connecting the Atlantic Ocean with a great body of water of which only the eastern banks are shown. It is apparently the hoped-for passage to the South Sea or Verrazzano's Bay, that westward strait or river firmly believed in by the Spanish, French, and English but not yet discovered. The English did not know that an expedition under Captain Juan Pardo sent by Pedro Menéndez de Avilés from Santa Elena had searched unsuccessfully for a water route. The most obvious source for White's unnamed strait at Port Royal is the Harleian (or Dauphin's) great wall map, ca. 1547, on which the waterway is called "R. de s.ᵉ helene" (plate 12). White's strait is longer, and his islands at Port Royal are based on le Moyne's map;[193] but similarities to the Harleian map are strong. There are other maps that White, Ralegh, and the planners of the Roanoke voyages may have known. The Gilbert-Dee ca. 1582 polar chart (plate 20) has at the same location a strait leading to the Verrazzano sea. An unsigned, undated chart (presently in the Greenwich National Maritime Museum, Greenwich, England), identified by the French cartographer Marcel Destombes as having been drawn by Juan Verrazzano, has the Verrazzano sea touching the Atlantic coast at "C. de dea elena."[194]

From Port Royal south to the Cape of Florida, White followed a map, no longer extant, by Jacques le Moyne de Morgues that showed the discoveries and nomenclature of the French. This map differed in numerous details from an engraved map of French Florida—based on another nonextant map by le Moyne that incorporated revisions of his own and from unknown sources—that de Bry published in 1591 in the second volume of his *America*, which included le Moyne's commentaries on the drawings. It is difficult to determine what changes de Bry made as engraver and how much variation existed between le Moyne's map, which White used, and that purchased by de Bry in 1588. A cursory comparison will reveal the principal differences: de Bry Latinized the names;[195] the locations of the lakes differ; and the engraved map includes a number of legends accompanying physical features such as mountains and lakes, which became standard features copied by European map makers for more than eighty years.[196]

White's version of French Florida portrays incorrectly the R. de May (St. John's River) as the effluent of Okefenokee Swamp; St. Mary's River, not shown, is the actual outlet. On the de Bry engraving the May correctly makes a left-angle turn south to "Lacus aquae dulcis" (Lake George). Both maps show Lake Okeechobee (White: "Secrope"; de Bry: "Serrope") in south Florida; but de Bry shows no outlet, while White correctly shows a river flowing west into the Gulf of Mexico.[197] On the Gulf coast of the Florida peninsula are a number of Spanish names that, except for "Baye de sa Ponce" (here Charlotte harbor, not Tampa Bay), are different from any Spanish maps known to be in England, as well as from the de Bry engraving. The French "baye" spelling indicates that it was, however, taken by White from le Moyne's map, not directly from some unknown Spanish source.[198] The shape of the Florida peninsula is unlike that shown on the engraved le Moyne map, which is almost an equilateral triangle, but similar to the Fernándes-Dee 1580 or the Dee 1580 in its narrow, almost isosceles triangular shape.[199] The Bahama islands are closer in shape, in the surrounding dotted (shallow) area and in the number and location of the smaller islands, to the Fernándes-Dee 1580 chart than to any other map examined. The similarities strengthen the great probability that Fernándes had the original map with him in 1585.[200]

The attempt by White to put together and reconcile the various maps and information that he combined on one general chart was not successful in its overall construction. He exaggerated the distance between the Cape of Florida and Port Royal—a distance based on le Moyne's map of French Florida—by more than 140 miles and the distance from Cape Lookout to Cape Henry by more than 215 miles. Even though White placed Cape Henry (Chesapeake Bay) at 40 degrees 45 minutes north latitude, almost three degrees too far north, he located Port Royal at the same latitude as Cape Lookout. Thus he solved the problem raised by the disproportionate size of Florida and Virginia by drawing the coastline directly east-west instead of in its true northeast-southwest direction, which would have put the Chesapeake at the latitude of Cape Cod.[201] The erroneous latitudes given on the map were, as David Quinn remarks, certainly not inspired by Harriot;[202] nor did White follow the northeasterly contour of the shoreline between Santa Elena (Port Royal) and Trafalgar (Cape Lookout) generally found on the Spanish maps of the period. But it should be noted that many able geographers for 150 years after the Juan Vespucci 1526 chart made an elongated, equal-latitude coast between Santa Elena and Trafalgar.[203]

54

Virginea Pars [1586?] (plate 24)[204]

This manuscript map by John White includes the coast from slightly south of Cape Lookout to the entrance of Chesapeake Bay and from the North Carolina Outer Banks to the mouths of the Neuse, Roanoke, and James rivers (none of which is named on the map). It is the most authoritative geographic record left of the coastline, islands, and Indian villages found during the first colony.[205]

The various Spanish configurations and toponymy of the coast are at last discarded entirely. For much of the mathematical and surveying skill evidenced, as well as the celestial and magnetic observations required, Thomas Harriot's presence during the field trips was undoubtedly responsible. The map's production is the result of ground surveys and marine sightings; the bearing of landmarks to each other is unusually correct; and the placement of and relationship among islands, natural features, and villages are reasonably accurate for a map of its scale. It is, as David Quinn states, "the most careful detailed piece of cartography for any part of North America to be made in the sixteenth century."[206] Even the great cartographic achievements of the early seventeenth century—John Smith's Virginia (1612) and Samuel de Champlain's St. Lawrence basin (1607, 1613, 1632)[207]—do not exhibit the careful ground surveying evident in "Virginea Pars."

The general correctness of outline and areal proportion of land bordering the Pamlico and Albemarle sounds—an area of about 3,000 square miles—would require, according to historian R. A. Skelton, careful reduction from a series of field notes similar to those Thomas Bavin was directed to take on large royal sheets in 1582, although no such field notes have survived.[208] Absent from White's map is much of the information called for in Bavin's instructions: depth soundings, rocks and other marine hazards, compass variations, differences in land formations, and native buildings. Either White and Harriot did not record such items cartographically or they were omitted because of the small scale of this particular map. There is little question that the two men collaborated, though only White made the final watercolor map. Harriot, as Ralegh's mathematician and the inventor of an improved backstaff and complex optical diagrams, was equipped for the topographical survey; and White, the skilled and observant artist, was the actual designer and creator of the maps. It may be that, as they explored the sounds, Harriot, with the aid of an assistant, observed the landmarks and their directions from each other and calculated their distances, while White made notes concerning what Harriot observed.[209]

White's "Virginea Pars" is a professionally drawn map with lightly shaded lines that distinguish the coast and the islands but with a minimum use of symbols. Large dots with identifying names, for example, rather than profiles or stockades, indicate native villages. The artist is not entirely absent: Indian canoes or dugouts are scattered in the sounds and up the rivers, two large vessels sail southward from Ocracoke

55

and Cape Lookout, and another ship with sails furled is anchored off Port Ferdinando. Explorations have disillusioned the colonists: there is no evidence of Roanoke River leading to the South Sea, to gold mines, or to Pamlico Sound being island-studded St. Mary's Bay.

The topography of the regions away from Pamlico and Albemarle sounds becomes markedly less accurate. The area within the barrier islands around Cape Lookout incorrectly expands as a wide, vaguely conceived extension of Pamlico Sound. White shows the important discovery of Chesapeake Bay, but his portrayal is geographically limited and confusing. It is a comparatively small open bay, with no arm extending northward to indicate recognition that it might be the Baya de Santa Maria or the Rio Salado of the Spanish charts. White's bay is bounded on the north by a straight shoreline; two rivers—the James and the York—extend westward; a large island—possibly a misplacement of Old Point Comfort—is shown in the bay near the mouth of the James. Its origin is clearly the island surrounded by deep water in the bay that Menatonon, the Chowanoac chief, described to Ralph Lane. The chief locations on the bay definitely referred to in the records are the villages of Chesepiuc on upper Lynnhaven Bay and Skicóac on the Elizabeth River, within the present limits of Norfolk.

Governor Lane sent a troop under a "colonel" for the winter of 1585-1586 to make allies of the "Chesepian" tribe and to search out the land. The troop probably went by sea or up Currituck Sound and by its tributary, North Landing River, inasmuch as Lane did not learn of the route by the Chowan River from Menatonon until March, 1586.[210] The governor must have received reports quite favorable for moving the colony from Roanoke, and White must also have received geographic information from the company wintering with the Chesepiuc-Skicóac Indians. He or Harriot or both of them may have visited the area themselves, although lack of identifiable drawings by White or commentaries by Harriot would seem to preclude any but a brief (as well as unreported) expedition. The paucity and inaccuracy of geographic knowledge concerning the bay indicate reliance upon second-hand information rather than surveying or exploration.

White-de Bry. Americae pars, Nunc Virginia dicta. 1590 (plate 25)[211]

When Theodor de Bry, the Frankfort publisher and engraver, decided to publish a collection of narratives about the New World, his choice for the first volume was Thomas Harriot's *briefe and true report of the new found land of Virginia,* to which he added a map and drawings by John White, as well as notes on them written in Latin by Harriot.[212] The map became the type or mother map for European map makers of the geographic delineation of the area for eighty years and longer.

The de Bry volume, which appeared almost simultaneously in four languages, obtained wide distribution, and the map it included quickly influenced European geographers. The map engraved by de Bry is very similar to White's "Virginea Pars," covering the area between Cape Lookout and Chesapeake Bay; the two maps exhibit sufficient differences, however, to posit a later revision by White or at least a different version as de Bry's actual and immediate source.

The engraved map extends farther inland to the sources of the Chowan, Neuse, and other estuaries of the sounds; the Roanoke reaches the mountains. De Bry added two inlets to White's eight, both in Currituck Sound. One, "Trinety harbor," was mentioned by Lane, who assigned it a depth of eight feet over the bar at high tide; the other, north of Croatamung Island, was at or near Mosketo Inlet, which remained open during most of the ensuing century. De Bry added sixteen new names to the map; White's "Virginea Pars" has more Indian settlements on the north side of Albemarle Sound, de Bry has more on the south. The engraved map makes much heavier use of symbols: palisaded villages and several figures, including a native warrior with bow and arrow; a great mountain range; shoals; and even symbols for varieties of trees. White may have added these symbols from more detailed plans or notes. De Bry also shows professional engraving characteristics. David Quinn comments that White's manuscript should take precedence over the engraving, especially in the location of villages, but that the changes are probably genuine alterations by the artist.[213] For the Chesapeake Bay area, Lynnhaven Bay is more detailed and more accurately drawn in the de Bry 1590 engraving than in the manuscript maps, and it forks correctly near its southern head. The Chesepiooc Indian settlement has been replaced near the west bank about fourteen miles below the entrance; a new Indian settlement, Apasus, is located near the entrance to Lynnhaven Bay. Only one Indian name remains on the north shore of Chesapeake Bay, which shows no knowledge of any extension stretching northward. Quinn notes that two islands off Cape Henry, apparently long since eroded, indicate observation of that area and may suggest that the Chesapeake expedition was made by sea. The cape would be an obvious place for lookouts to watch for vessels, friendly or hostile, and could also be reached by land.

White-de Bry. The arriual of the Englishemen in Virginia. 1590 (plate 26)[214]

Theodor de Bry also engraved a semipictorial map of a kind very popular at the time that includes the coast from Cape Kendrick (present Wimble Shoals) to Currituck Inlet (the present North Carolina-Virginia line). Outside the barrier islands are two large vessels riding at anchor; at each of the five inlets shown is a sunken ship, probably indicating the dangers of entrance rather than actual shipwrecks. From "Trinety harbor" inlet a pinnace is approaching Roanoke Island, where there is a palisaded Indian village. White's pictorial map is much richer in symbols than his manuscript maps. It shows half a dozen varieties of trees, two dripping with Spanish moss; grapevines where the sketch map had indicated their presence; a

57

fish weir; and several palisaded Indian villages. Shoals extend across Albemarle Sound. Usually the occupant of the rear of a dugout sits, while the poler in front stands; on Roanoke Island an Indian with drawn bow is aiming at an antlered deer. No original drawing corresponding to this one is found in the White collection; it overlaps the northern portion of the sketch map sent back to England about September, 1585, which covers nearly the entire barrier island chain, although on a smaller scale and more simply. White may already have seen a pictorial map or view of the arrival of the French at Port Royal in 1562, drawn by Jacques le Moyne, the fellow painter in Ralegh's employ;[215] this work, along with several other pictorial views of the Georgia-South Carolina coast, was purchased by de Bry from le Moyne's widow in 1588 and engraved for the second volume of his *America* series. It cannot compare in accuracy or authenticity with White's "arriual," inasmuch as le Moyne did not actually accompany Jean Ribault on his first voyage of 1562. White's pictorial map, for all of its accuracy of detail, fails, however, to correspond exactly with the accounts of either the arrival in 1584 of Amadas and Barlowe or the arrival in 1585 of the first colony.[216]

Some Post-Roanoke Maps

The only sixteenth-century cartographic production that adds new information to the White and White-de Bry maps is the great globe designed between 1589 and 1592 by the English geographer Emery Molyneux with the assistance of the mathematician Edward Wright in London under the patronage of William Sanderson, nephew by marriage of Sir Walter Ralegh and wealthy supporter of his Virginia ventures. The globe was engraved by Jodocus Hondius in 1592. No geographic additions are found on the globe, but variations in the spelling of names and the inclusion of new names indicate the use of published reports, as well as consultation with someone who had been in Virginia. "Endesokee," a barrier island below Wococon, and "Cape of Feare" (Cape Lookout; de Bry's "Promontorium Tremendum") appear for the first time.[217]

Even before de Bry's volumes appeared, a few maps that showed the English colony were being produced. Richard Hakluyt the younger, ever the promoter of English overseas expansion, furnished the information for *Novvs Orbis*, a finely drawn map based largely on some Spanish-type original. *Novvs Orbis* was made by the Antwerp engraver Filips Galle in 1587; it accompanied Hakluyt's edition of Peter Martyr's *De orbe novo . . . decades octo.* The map has "Virginea 1584" above a long bay or coastal indentation with many islands extending from 35 to 40 degrees north latitude, located with possibly intentional vagueness. On the map are other legends concerning voyages of English discoveries: Sebastian Cabot's Bacallaos 1496 (Labrador), Martin Frobisher's Meta incognita 1576 ("the unknown Goal," here Baffin Island), Francis Drake's Nova Albion 1580 (California), and other geographic additions.[218] Baptista Boazio's "The Famouse West Indian voyadge" (1588?) shows the "pricked line" of Drake's return voyage of 1586, during which he retrieved the first Roanoke colony.[219] Franz Hogenberg, a famous engraver of Cologne, created *Americae Descriptio,* a map of the Western Hemisphere, in 1589; the map shows "Virgine" as a city off the coast and slightly above the fortieth parallel.[220]

The well-entrenched Spanish padrón concept of the east coast of America remained on many new maps and in many new atlases, both manuscript and printed; for those map makers influenced by Spanish or Portuguese geographers, the older delineation and nomenclature appeared more authoritative and easier to follow. De Bry's engravings of White's "Virginia" (plate 25) and le Moyne's "Florida," however, made available to enterprising cartographers remarkably detailed portions of the coast that they were not slow to utilize, although the intervening extension of coast between the two maps caused confusion and, on some maps, an erroneous interpolation of Spanish geographic information that lasted for many decades.

In 1593 there appeared a large map of North America that used both maps for the southeast coast of the continent in the notable atlas *Speculvm Orbis Terrae,* by the Dutch cartographer Cornelius de Jode. Unfortunately de Jode inserted between le Moyne and White the entire coast from Florida to New England, as shown on the old Spanish maps; thus White's Chesapeake Bay was raised to the latitude of Boston. De Jode compensated for this error by emphasizing still further the east-west coastline for New England.[221] In 1597 the Dutch geographer Corneille Wytfliet published two maps. One, of Florida and Apalche, utilized le Moyne's "Florida" and a Spanish-type map by the Dutch cartographer Abraham Ortelius as a basis; it extends northward to C. de Arenas (Cape Cod, at 38½ degrees north latitude). In the second, of Norumbega (New England-Newfoundland) and Virginia, the lower portion is based on White's 1590 map, but all the White details are inserted above C. de Arenas, placing Chesapeake Bay at 43 degrees north latitude. These two maps appeared in Wytfliet's *Descriptionis Ptolomaicae Augmentvm* (1597), which is sometimes called the first atlas of the New World inasmuch as it contains no maps except those of the Americas.[222] The work was highly regarded and followed by many geographers: José de Acosta, a Spanish historian, in 1598 and Jean Matal, a German, in 1600 each published two maps, both apparently smaller copies of Wytfliet's.[223]

De Bry's maps of Virginia by White and of Florida by le Moyne altered the general European perception of the entire southeastern North American coast—not directly so much as through a later adaptation of one map that became one of the important cartographic works of the early colonial period. In 1606 Jodocus Hondius, a Flemish cartographer of Ghent who had moved to Amsterdam about 1593 after having worked at his craft for a decade in London, designed and executed "A New Description of the American Provinces of Virginia and Florida."[224]

59

(plate 27) The Hondius map is regarded as a splendid example of the cartographic art. Its ornamental title cartouche includes two side panels with Indian village scenes from White's paintings and le Moyne's engravings, and elsewhere are vignettes of inhabitants of the region. The bold swashes in lettering that are characteristic of Amsterdam map makers in this golden age of Dutch cartography and that serve to fill up the unknown spaces of the interior add greatly to the effect, if not to the accuracy, of the map.

In London, more than a dozen years earlier, Hondius had already used White's map when engraving the gores (map sections) for the Molyneux globe. In 1603 he bought the plates of Gerhardus Mercator's *Atlas* and made many new plates for his first 1606 Latin edition; he eventually published it in four languages and numerous editions, although he preserved Mercator's name with its reputation on the title page.

Hondius followed White-de Bry (plate 26) in making Chesapeake Bay a small indentation on the coast. He also followed most of White's physical details of the Outer Banks area, the rivers, and the names of the Indian villages. He made a number of changes, however, that distort the coastline. He increased the distance from Albemarle Sound to Chesapeake Bay (nothing is shown from the Chesapeake to Cape Cod) and lengthened the Outer Banks so that Cape Lookout is below the actual latitude of Port Royal.[225] He then overcompensated for this by extending the coastline west by north instead of southwest, consequently placing Port Royal half a degree north of its actual latitude. In truth, little or nothing was known of the interior country of present North Carolina west of the coast, but what Hondius placed there was, for the average European, the most exciting part of the map. He used only the northern portion of le Moyne's *Florida* above the peninsula but transferred details from it to the "Apalatcy" (Appalachian) mountains, called them "gold-bearing," and noted that "the natives find grains of silver in a lake fed by a great waterfall." Farther west is a lake of water so great that "one is unable to see from one bank across to the other shore," a tantalizing suggestion of a possible route to the South Sea.[226] Depicting an area of contemporary geopolitical interest with exceptional artistic skill, this map continued to be copied by continental map makers such as W. J. Blaeu, Jan Jansson, Arnoldus Montanus, and others long after the improved second map of the Lords Proprietors of Carolina (Gascoyne 1682) appeared in the last quarter of the century.

One other map of the early seventeenth century should be mentioned—not so much because it derives from the cartography of the Roanoke colonists as for the information contained in its legends and the search that resulted from the employment of these legends. In September, 1608, the Spanish ambassador to England, Pedro de Zuñiga, obtained a map sent to England from the newly established colony at Jamestown and dispatched it or a copy of it to King Philip II.[227] This crudely drawn chart of the Chesapeake Bay region (plate 28) includes three rivers to the south of the James River that represent the Neuse, Pamlico, and Roanoke. The colonists had special orders to find out from the Indians where survivors of the 1587 colony might be, in order to rescue them. On the Zuñiga chart at Pakerakanick on the Neuse River is the following legend: "Here remayne the 4 men clothed that came from Roonock [Roanoke] Oconohowan." North of Pamlico River is another legend: "Here the king of paspahegh reported our men to be and went to se[e]." And on the south bank of the James this note: "Here paspahegh and 2 of our men landed to go to panawiock." The search for survivors was never, for whatever cause, successful.

The Jamestown colonists were also ordered to search for gold and for a passage to the South Sea. Neither was found; but on the Zuñiga chart, to the west, near the head of the James River, near a seacoast, is this legend: "Hear [sic] the salt water beatethe into the riuer amongst theis rocks being the south sea." The legend is reminiscent of Governor Ralph Lane's report from the chief Menatonon and other Indians that the source of the Roanoke River was a great spring rushing out of a "hugh rocke [that] standeth nere vnto a Sea. The waves thereof are beaten into said fresh streame."[228] The stream rushing out of a rock probably had its origin in stories of the waterfalls in the Allegheny Mountains, where the coastal rivers have their source. The nearness of the Pacific Ocean beyond the Blue Ridge Mountains remained a firmly held belief until the end of the seventeenth century among many Virginia officials in England and America.

IX. Conclusion

This book is but a part of the examination of an important episode: the first attempted English settlement in America. Historical interpretation, documentary research, archaeological investigation, imaginative re-creation in the arts—all are making their contributions to a more complete understanding of what occurred. What has the study of cartography to add to this pursuit of knowledge?

Early maps are among the most valuable historical documents when one learns to read them correctly. They show what was known or believed about the land at different periods and the extent and limitations of knowledge at various times in the past. It is often as important to ascertain what was known or believed wrongly as rightly; such misjudgment explains actions that would otherwise be unintelligible. Early maps also show significant changes in geography that without them would not be known.

This book has attempted to show that cartography has indeed a contribution to make to the understanding of the Ralegh voyages and the people who made them—their thinking, their purposes, their plans, their choice of location, their misapprehensions, their actions, their disasters, and their accomplishments. When one enters a new territory of any sort in life, the first need is for some sort of map. And when one departs from it, the best legacy one can leave to those who follow is, in its turn, a better map.

What maps were or could have been available to the planners of the Ralegh voyages, and how did they influence their thinking? Maps of new lands spring from exploration; thus, one must consider the matter of who had explored the southeastern coast of North America. There was Giovanni da Verrazzano with his illusory isthmus and great South Sea bay that spurred the fruitless search of persistent explorers for nearly two centuries. The Spanish maps of the Casa de Contratación, based on the voyages of the Licenciado Lucas Vasquez de Ayllón, Pedro de Quexós, and Estéban Gómez, misled the English into thinking that a broad bay, studded with protecting islands, would welcome them where the Outer Banks, one of the most dangerous coasts in the world, extended its harborless length. The French Huguenots and the Dieppe school of map makers furnished maps; but even more vividly the Spanish massacre of the French colonists on the St. John's River in 1565 demonstrated the geographic proximity of the Spanish and that government's determination to destroy intruders. Richard Hakluyt the younger, fervent promoter of colonization and prolific publisher of accounts of earlier explorers, printed maps and referred in his writings to others that were available. Contemporary explorers and navigators, such as the English colonists' widely experienced pilot and forceful leader Simon Fernándes, with his burning ambition to capture Spanish treasure ships passing near this coast, had their maps. A supply of maps of varying quality came into English hands from foreign vessels captured by privateers that scoured the Spanish main.

With the arrival of the English colonists on the coast came the harsh geographic realities, which were vastly different from their expectations. The colonists' adaptation to their surroundings was indeed difficult; it was not lessened by the frustration of a colony poorly supported by supplies from home. Geography as portrayed by Giovanni da Verrazzano, with its representation of the Pacific Ocean so tantalizingly near, its visual assertion on many maps that from this forbidding coast the sea route to Cathay could be, as Hakluyt wrote, "searched out as well by River and overlande as by sea," had made a deep impression on the colonists' minds. They clung to the latitude judged best to begin the search, which ended repeatedly in failure. The great open bay with its sheltering islands, promised them at this latitude by the authoritative maps of the Casa de Contratación and their adaptations from the trade center of Dieppe, was to have been their home. They found instead a storm-beaten barrier with inlets so treacherous that only their smallest ships could pass through and shallow, landlocked sounds surrounding their little island refuge.

It is strange that even after the English colonists reached the Outer Banks at the latitude where they had so often found the great bay on Spanish-type maps, they did not appear to realize that Pamlico Sound was not the Baya de Santa Maria, nor Albemarle Sound and the Roanoke River the Rio Salado, the route to gold and silver mines and to the Pacific. And even after they had visited the shore of Chesapeake Bay, there is no clear indication that they identified it with the Baya de Santa Maria.

Nevertheless, out of Sir Walter Ralegh's colonial ventures came a cartographic accomplishment that emanated from the conjunction of the genius of the painter John White with that of the scientist Thomas Harriot. The surveys conducted by these men resulted in a successfully accurate mapping of the region they had explored. The publication of their map made a permanent and widely available record that for the first time presented to Europeans reasonably accurate and usable knowledge of that part of the New World. White and Harriot did indeed, in their turn, leave a better map.

Notes

[1]David Beers Quinn (ed.), *The Roanoke Voyages, 1584-1590* (Cambridge, England: Hakluyt Society, 2 volumes, 1955), I, 91-92, hereinafter cited as Quinn, *The Roanoke Voyages.*

[2]In spite of the legend on the map, "Juan de la cosa la fizo en el puerto de S: Mjª en año de 1500" (John de la Cosa made it at the Port of Santa Maria in the year 1500), questions concerning the map, including its authorship and date, have caused much dispute. For a basic discussion and bibliographies, see W. F. Ganong, *Crucial Maps in the Early Cartography and Place-Nomenclature of the Atlantic Coast of Canada.* Introduction, commentary, and map notes by Theodore E. Layng (Toronto: University of Toronto Press/Royal Society of Canada, 1964), 8-43, hereinafter cited as Ganong, *Crucial Maps of the Atlantic Coast of Canada;* and the essay by R. A. Skelton entitled "The Cartography of the Voyages," in James A. Williamson, *The Cabot Voyages and Bristol Discovery under Henry VII* (Cambridge: Hakluyt Society, second series, 1962), 120, 295-307, hereinafter cited as Skelton, "The Cartography of the Voyages."

[3]The Cantino, Waldseemüller, and related maps have, like the Cosa, been the subject of much critical examination. See bibliographies in T. E. Layng (ed.), *Sixteenth-Century Maps Relating to Canada* (Ottawa: Public Archives of Canada, 1956), 6 ff, (hereinafter cited as Layng, *Sixteenth-Century Maps Relating to Canada,*) and Lawrence C. Wroth, *The Voyages of Giovanni Verrazzano, 1524-1528* (New Haven: Yale University Press for the Pierpont Morgan Library, 1970), 44-53, hereinafter cited as Wroth, *The Voyages of Verrazzano.*

The name America was first given to the southern continent of the New World in 1507 in the *Cosmographiae Introductio* and on two maps issued simultaneously with it: the great world map *Universalis Cosmographia* (on twelve sheets, each 18.2 inches by 24.8 inches) and a smaller sheet (9½ inches by 15 inches) of connected segmented gores for a globe. All three were ascribed to Martin Waldseemüller, the learned cartographer and monk of St. Dié, by Alexander von Humboldt in the 1830s, as well as by many subsequent scholars. Reexamination of the documents indicates clearly, however, that the *Introductio* was written by Matthias Ringmann ("Philesius"), who included in the text the self-laudatory accounts of four voyages to the New World by Amerigo Vespucci, later justly questioned as to accuracy. (Franz Laubenberger, "The Naming of America," *Sixteenth Century Journal,* XIII [1982], 91-113.) Ringmann's text proposed that the continent be named "Amerigen" or "America" and apparently persuaded Waldseemüller to enter the name on his map and gores. But after the death of Ringmann in 1511, Waldseemüller's maps (Ptolemy 1513, on which the cartographer correctly attributes the discovery to Columbus, and the great Carta Marina 1516) omit the name. The influence of the *Introductio* and the world map of 1507 was pervasive, however, and "America" was soon extended to include both continents.

[4]Belief, however, in a Central American strait to the Pacific somewhere south of Yucatan persisted until the middle of the sixteenth century. Vesconte de Maggiolo's 1527 map shows a "stretto dubitoso" below Yucatan; his 1549 map shows it without a "dubitoso." Verrazzano was searching for a strait near Darien during his third voyage in 1528 when he was seized and eaten by cannibals. See Wroth, *The Voyages of Verrazzano,* 248-249.

[5]James Alexander Robertson, *Magellan's Voyage around the World by Antonio Pigafetta* (Cleveland, Ohio: N.p., 3 volumes, 1906), II, 189, 272, hereinafter cited as Robertson, *Magellan's Voyage around the World;* Wroth, *The Voyages of Verrazzano,* 220.

[6]The definitive work on Verrazzano's voyages is Wroth, *The Voyages of Verrazzano.*

The Guadagni family of Florence commissioned, or was presented with, the Agnese atlas, presently in B. L. Egerton 2884, which contains a fine map of the North American coast.

Francis I's direct involvement in and support of Verrazzano's voyage is shown conclusively in six letters of 1523 to Vice Admiral de Chillon, commander of the channel ports, signed by Francis. These letters, sold by Stevens to the Research Library of the University of California at Los Angeles, have been printed and annotated in Norman J. W. Thrower, "New Light on the 1524 Voyage of Verrazzano,"*Terrae Incognitae,* XI (1979), 59-65.

[7]Wroth, *The Voyages of Verrazzano,* 135.

[8]Wroth, *The Voyages of Verrazzano,* 135.

[9]Wroth, *The Voyages of Verrazzano,* 136 n. 8 (Verrazzano's marginal note). Wroth's work on Verrazzano has a facsimile of the Cèllere Codex, a transcription, and a translation from the Italian by Susan Tarrow.

[10]See plate 5.

[11]Wroth, *The Voyages of Verrazzano,* 136 n. 8 (Verrazzano's marginal note).

[12]Wroth, *The Voyages of Verrazzano,* 83, 137 n. 9 (Verrazzano's marginal note).

[13]Ganong, *Crucial Maps of the Atlantic Coast of Canada,* 111; Wroth, *The Voyages of Verrazzano,* 82-83; Samuel Eliot Morison, *The European Discovery of America: The Northern Voyages, A.D. 500-1600* (New York: Oxford University Press, 1971), 295-296, hereinafter cited as Morison, *The European Discovery of America.*

[14]The Maggiolo world map was destroyed in a bombing raid in World War II that hit the Biblioteca Ambrosiana in Milan, Italy; the Verrazzano is in the Vatican library in Rome. Both have been reproduced frequently. Edward Luther Stevenson (ed.), *Maps Illustrating Early Discovery and Exploration in America, 1502-1530* (New Brunswick, N.J.: twelve maps, 1903-1906, each in separate folder), hereinafter cited as Stevenson, *Maps Illustrating Early Discovery,* includes a full-scale photographic facsimile of each. Wroth's *Voyages of Verrazzano* (pp. 170-177, 296, figures 19-22) analyzes and lists all important Verrazzano-type maps, locates the best reproductions, and gives many itself. Ganong's *Crucial Maps of the Atlantic Coast of Canada* (pp. 169-174) has the most detailed analysis of place-names on these two maps, with a parallel list of all names on the North American coast in the Cèllere Codex and the maps and tracings of the maps in an end folder (figures 34-36). Other useful studies of the New England-Canadian section of the Verrazzano-type maps are also made by I. N. Phelps Stokes, *The Iconography of Manhattan Island, 1498-1909* (New York: Robert H. Dodd, 6 volumes, 1915-1928; reprint, New York: Arno Press, 1967), II (introductory essay by F. C. Wieder); and Bernard G. Hoffman, *Cabot to Cartier: Sources for a Historical Ethnology of Northeastern North America, 1497-1550* (Toronto: University of Toronto Press, 1961), 105-125, hereinafter cited as Hoffman, *Cabot to Cartier.*

[15]Treviso, Biblioteca Comunale, MS425, fol. 4, Giuseppi Caraci, "La Produzione Cartografica di Vesconte Maggiolo (1511-1549) ed il Nuovo Mondo," *Memorie Geografische,* IV (1958), 221-289, with repr., Tav. 5, opp. p. 282.

[16]David B. Quinn, *Richard Hakluyt, Editor: A Study Introductory to the Facsimile Editions of Richard Hakluyt's Divers Voyages (1582)* . . . (Amsterdam: Theatrvm Orbis Terrarvm, 2 volumes, 1967), II, [7], hereinafter cited as Quinn, *Richard Hakluyt, Editor;* and R. A. Skelton's essay entitled "Hakluyt's Maps," in David B. Quinn, *The Hakluyt Handbook* (London: Hakluyt Society, 1974), I, 48-73.

[17]The two maps differ markedly in latitudes for the entire North American continent: the Cape of Florida (25° N.L.) is 20° N.L. on the Maggiolo 1527 and 33° N.L. on the Verrazzano 1529.

[18] Richard Hakluyt, *The Principal Navigations Voyages Traffiques & Discoveries of the English Nation* (New York: AMS Press, 12 volumes, 1965), II, 2, 287, hereinwith cited as Hakluyt, *The Principal Navigations;* also reprinted in David B. Quinn (ed.), *New American World: A Documentary History of North America to 1612* (New York: Arno Press, 5 volumes, 1979), III, 106, hereinafter cited as Quinn, *New American World.*

[19]Wroth, *The Voyages of Verrazzano,* 165-216, has the best list of Verrazzano-influenced maps, along with critical analysis and numerous reproductions of maps and globes.

[20]Harold L. Ruland, "A Survey of the Double-page Maps in Thirty-five Editions of the *Cosmographia Universalis* (1544-1628) of Sebastian Münster, and in his Editions of Ptolemy's *Geographia* 1540-1552," *Imago Mundi*, XVI (1962), 84-97, with a census of both maps, hereinafter cited as Ruland, "A Survey of the Maps in the *Cosmographia Universalis*"; Wroth, *The Voyages of Verrazzano*, 191-192.

[21]William P. Cumming, *The Southeast in Early Maps: With an Annotated Check List of Printed and Manuscript Regional and Local Maps of Southeastern North America During the Colonial Period* (Princeton: Princeton University Press, 1958; second edition, Chapel Hill: University of North Carolina Press, 1962), 7, 65 n. 20, listing some of the variant spellings of the name on early maps, hereinafter cited as Cumming, *The Southeast in Early Maps*.

[22]Gilmary Shea wrote the account of Ayllón's enterprise in the chapter "Ancient Florida" in Justin Winsor (ed.), *Narrative and Critical History of America* (Boston: Houghton, Mifflin and Company, 8 volumes, 1884-1889), III, 231-298, hereinafter cited as Winsor, *Narrative and Critical History*, from documents he discovered as well as from other contemporary accounts. Henry Harrisse, *The Discovery of North America: A Critical, Documentary, and Historic Investigation, with an Essay on the Early Cartography of the New World* (London: H. Stevens, 1892), 198-213, hereinafter cited as Harrisse, *The Discovery of North America*; and Woodbury Lowery, *The Spanish Settlements within the Present Limits of the United States, 1513-1561* (New York and London: G. P. Putnam's Sons, 2 volumes, 1901, 1905), I, 153-168, hereinafter cited as Lowery, *The Spanish Settlements*, have added to the details of Spanish historians such as Peter Martyr and Oviedo, who knew the voyagers personally and had access to documents from archives and court records. Ayllón was a licentiate with a permit to practice law.

[23]Winyah Bay, South Carolina, into which several rivers flow, is the nearest "river" at 33°13'; the Black, Pee Dee, and Waccamaw rivers flow into it.

[24]Gonzalo Fernandez de Oviedo y Valdés, *Historia general y natural de las Indias, isles, y tierra-firme del mar oceano* (Madrid, Spain: N.p., 4 volumes, 1851-1855), lib. XXVII, cap. 1, Vol. III, p. 628, hereinafter cited as Oviedo, *Historia general y natural de las Indias*, quoted in Harrisse, *The Discovery of North America*, 212.

[25]Lowery, *The Spanish Settlements*, I, 165.

[26]Oviedo, *Historia general y natural de las Indias*, III, 628.

[27]Samuel Eliot Morison, *Admiral of the Ocean Sea: A Life of Christopher Columbus* (Boston: Little, Brown and Co., 1942), 191, estimated Columbus's land-coasting league was only 1½ miles, his sea league 3.18 nautical miles. However, Alonso de Chaves, pilot major, later professor of navigation of the Casa de Contratación, and maker of the 1536 padrón general, generally equated coasting distances as twenty leagues to one degree in the *derrotero* of his *Qvatripartitv[m]* (ca. 1539), equivalent to about 3½ miles to a league. Oviedo possessed and used a copy of Chaves's 1536 padrón general.

[28]For the Cape Fear River as the Jordan and Winyah Bay as Gualdape, see Lowery, *The Spanish Settlements*, I, 166, and his Appendix H, which gives the best listing and summary of the early dated documents; Paul Quattlebaum, *The Land Called Chicora: The Carolinas under Spanish Rule* (Gainesville: University of Florida, 1956), 20-31, and Appendix B, hereinafter cited as Quattlebaum, *The Land Called Chicora*, gives a strong but one-sided support of Winyah Bay as Gualdape. For the Savannah River, see John R. Swanton, *Early History of the Creek Indians and Their Neighbors*, Bulletin 37, United States Bureau of American Ethnology (Washington, D.C.: Government Printing Office, 1922), 36-37, 66-67, hereinafter cited as Swanton, *Early History of the Creek Indians*; for the James River, Virginia, area see John G. Shea, "Ancient Florida," in Winsor, *Narrative and Critical History*, II, 241, hereinafter cited as Shea, "Ancient Florida." Chaves, the pilot major, who from personal interviews with survivors should have in his *Qvatripartitv[m]* an almost authoritative voice, apparently confused the problem; he wrote (fol. 43): "River Jordan 33½° northeast of Cape Santa Elena 30 leagues. West of Cape Sant Roman 4 leagues. This is a large [grande] river. This river calls itself River of aillon because he died there." Winyah Bay is 30 leagues along the coast northeast of Cape Santa Elena. See the coastal nomenclature of Chaves in the appendix.

[29]Edward Gaylord Bourne (ed.), *Narratives of the Career of Hernando De Soto* (New York: A. S. Barnes and Co., 2 volumes, 1904), 11, 13, 14, hereinafter cited as Bourne, *Narratives of Hernando De Soto.*

[30]Frederick W. Hodge and Thomas H. Lewis, *Spanish Explorers in the Southern United States, 1528-1543* (New York: Scribner, 1907), 174, hereinafter cited as Hodge and Lewis, *Spanish Explorers.*

[31]Harrisse, *The Discovery of North America,* 744-745.

[32]Vespucci may well have known Ayllón personally and heard from him of his planned colony in 1523, when Ayllón was in Spain securing his grant. Peter Martyr, the hospitable historian of the *decades,* often had in his home Ayllón and the imaginative Indian storyteller Francisco of Chicora, whom Quexós had captured during his 1521 voyage. Of Vespucci, Peter Martyr wrote: "Vesputius is my verye familyar frende, and a wittie younge man in whose company I take great pleasure, and therefore vse hym often tymes for my geste." See Harrisse, *The Discovery of North America,* 745, quoting Hakluyt's translation of Martyr, *decades* (London: Henry Stevens, 1892), Part III, 215.

[33]Port Royal was so named by Ribault during his exploratory voyage along the coast in 1562. Early explorers along a relatively unknown coast usually gave their own names to landmarks, especially if they were from a different country; this is, of course, an additional cause of nomenclatural confusion. In the 1660s the Massachusetts adventurer William Hilton, seeking a good place for a settlement, reached Cabo de Santa Elena and called it Hilton Head. The Spanish Rio de Santa Elena, four leagues up the coast, was not renamed; its entrance remains St. Helena Sound, although the river is the South Edisto.

[34]For a reproduction in color of the complete Vespucci world map, see William P. Cumming, R. A. Skelton, and David Beers Quinn, *The Discovery of North America* (London: Elek Books, 1971; New York: American Heritage Press, 1972), 86-87, hereinafter cited as Cumming, Skelton, and Quinn, *The Discovery of North America.* See Cumming, *The Southeast in Early Maps,* plate 2, for the detail of the southeastern North American continent. The coast turns sharply east at Cabo de Santa Elena (41°; the actual latitude of Hilton Head is 32° 14') and continues east at the same latitude to Cabo de Santa Maria (actually about 37°).

[35]L. A. Vigneras, "The Voyage of Estéban Gómez from Florida to the Bacalaos," *Terrae Incognitae,* II (1970), 28, hereinafter cited as Vigneras, "The Voyage of Estéban Gómez." Professor Vigneras found new documents that deal with preparations for the voyage and activities in Coruña following Gómez's return. He has effectively disproved the older theory that Gómez sailed directly to New England and followed the coast down to Cuba.

[36]For reproductions, see Cumming, Skelton, and Quinn, *The Discovery of North America,* 74; Quinn, *New American World,* I, plate 45; Amando Cortesão, "Note on the Castiglioni Planisphere," *Imago Mundi,* XI (1954), 53, hereinafter cited as Cortesão, "Note on the Castiglioni Planisphere."

[37]For a parallel table of Gómez names on the New England coast on the Ribero-type maps, see Ganong, *Crucial Maps of the Atlantic Coast of Canada,* 150-151; see also William P. Cumming, "The Colonial Charting of the Massachusetts Coast," in *Seafaring in Colonial Massachusetts* (Boston: Colonial Society of Massachusetts, 1980), Appendix B, "The Gómez Nomenclature," hereinafter cited as Cumming, "The Colonial Charting of the Massachusetts Coast." Two other unsigned maps, made shortly after the Castiglioni map, are now attributed to Ribero. The Salviati map (ca. 1526) adds six more names to the New England coast and has on the "Ayllón coast" Rio de las Osteas (32°), aguada (33°; "watering place"), b. delas canas (34½°), Costa baxa (37°; "shoal coast"). Three of these are descriptive; the two place-names are not repeated on any later maps. For a Salviati reproduction, see Cumming, Skelton, and Quinn, *The Discovery of North America,* 72, and Quinn, *New American World,* I, plate 46. The Weimar-Spanish (1527) map, possibly copied in Seville from a Ribero original made in Coruña, has only "aguada," "Tierra del licenciado ayllon," and "R. de S. Juhan." These names are copied by the writer from the originals.

[38]Harrisse, *The Discovery of North America,* 261-266; Edward L. Stevenson, "The Geographical Activities of the Casa de la Contratación," *Annals of the Association of American Geographers,* XVII (June, 1927), 39-59, hereinafter cited as Stevenson, "The Geographical Activities of the Casa de la Contratación."

[39]L. A. Vigneras, "The Cartographer Diogo Ribeiro," *Imago Mundi,* XVI (1962), 80-81, hereinafter cited as Vigneras, "Diego Ribeiro."

[40]Stevenson, "The Geographical Activities of the Casa de la Contratación," 42-43. Presumably, only such parts of the padrón as were needed for the intended voyage or mission were copied. Compare with Vigneras, "Diego Ribeiro," 79.

[41]Harrisse, *The Discovery of North America,* 632. The Seville padrón must have been at least the size of the master chart kept at Dieppe, which Dr. Helen Wallis suggests, from the size of the adjoined sheets in an atlas presumably based on it, might have been about 13 feet wide. See the introduction to Johne Rotz, *The Maps and Text of the Boke of Idrography Presented by Johne Rotz to Henry VIII.* Edited by Helen Wallis, with a foreword by Viscount Eccles (Oxford: Printed for Presentation to the Members of the Roxburghe Club, 1951), hereinafter cited as Rotz, *The Maps and Text of the Boke of Idrography.* Wallis's introductory essay to the Rotz volume is one of the most thorough and informative studies of the sources, construction, and relationships of a sixteenth-century world map that has been made.

[42]Harrisse, *The Discovery of North America,* 263-267; Stevenson, "The Geographical Activities of the Casa de la Contratación," 48-50.

[43]Oviedo, *Historia general y natural de las Indias,* III, 627 (Prohemio); Lowery, *The Spanish Settlements,* I, 447.

[44]Vigneras, "Diego Ribeiro," 80-82.

[45]Harrisse, *The Discovery of North America,* 557; Ganong, *Crucial Maps of the Atlantic Coast of Canada,* 145-146.

[46]Vigneras, "Diego Ribeiro," 82.

[47]See the comparative list of southeastern coastal names on Spanish maps in the appendix; for a similar list of the Gómez names on the New England coast, see Ganong, *Crucial Maps of the Atlantic Coast of Canada,* 150-151. For a color reproduction of the Ribero-Vatican 1529 planisphere, see Cumming, Skelton, and Quinn, *The Discovery of North America,* 106-107. The best reproduction of the Ribero-Weimar 1529 is the photograph in Stevenson, *Maps Illustrating Early Discovery,* No. 11. The best list of various reproductions for the printed and manuscript maps of this century is in Layng, *Sixteenth-Century Maps Relating to Canada.* Vigneras, "Diego Ribeiro," 81, places the drafting of Ribero-Weimer 1529 after that of Ribero-Vatican 1529 because of its greater number of legends (52 vs. 17) and other details; but Ganong, *Crucial Maps of the Atlantic Coast of Canada,* 147-150, points out numerous additions in names on the Vatican map, more correctness in their spelling and location, and other "improvements" over the Weimar manuscript.

[48]Harrisse, *The Discovery of North America,* 631-636.

[49]*Qvatri-partitv en cosmographia pratica . . . por Alonso de Chaves cosmographo Dela Magestad,* Biblioteca, Real Academia, Madrid, 9-14-1, 2791, hereinafter cited as *Qvatri-partitv en cosmographia pratica.* Chapter XIII, fols. 41v-44r, cover the coast from Florida to Cape Cod. The names are given in the appendix. For a valuable analysis of the work, see Ursula Lamb, "The Quarti Partitu en Cosmographia by Alonso de Chaves. An Interpretation," *Agruparmento de Estudos de Cartografia Antiga,* XXVIII (Coimbra, 1969), 3-9 (offprint of *Revista do Universidade de Coimbra,* Vol. XXIV).

[50]The description by Chaves (fols. 42v-49) makes the identification of these two unmistakable. Trafalgar, at 35½°, is 14 leagues directly east of Rio de principe and 20 leagues due south of Baya de Santa Maria at 36½°. This also clarifies the length of a league, since 1° is very nearly 70 miles, and a league, therefore, is approximately 3½ miles. This equivalency is reinforced by other north-south measurements in Chaves's *derrotero.*

Cabo de Santiago (Cape Cod), at 40°, the most prominent cape on the north coast, is 30 leagues north of Cabo delas Arenas (Monomoy Point) at 38⅓°; Cabo de Santiago forms the eastern shore of Baya de Cristoual, a great bay at 40° that extends west of it. No landmark on the coast conforms to these directional specifications, with the exception of Cape Lookout (with a coast to the west of it on the same latitude) and Cape Cod (with a large bay to its west).

⁵¹Chaves may have gained additional information from other later voyagers along the coast, who described the Chesapeake in more extensive detail. His description of a great bay with many islands is still ambiguous: the lower Chesapeake is uncluttered with islands, but it and the maps following the 1536 padrón approximate more closely the lower Chesapeake than the Ribero-type open bay with its double row of islands.

The Chaves description (fol. 44) follows: "The Baya de santa maria in the northern coast [the southern coast ends with Trafalgar] is in 36½ degrees. It is to the north of the Cape of Trafalgar [Cape Lookout] 20 leagues. It is southwest of the Cape of Sant Juan [Cape Henlopen?], twenty leagues.

It is a great bay with many islands in it. Two rivers flow into it; the one Espiritu Santo and the other Rio Salado [Salt River]. The Rio de espiritu santo is in 36½ degrees. This enters the bay of Santa Maria to the Southwest of it. The Rio Salado is in 37 degrees. This enters the Bay of Santa Maria to the north of it."

⁵²Rotz, *The Maps and Text of the Boke of Idrography,* 40, quoting Henry Harrisse, *Jean et Sebastian Cabot* (Paris: Leroux, 1882), 72 n. 3, hereinafter cited as Harrisse, *Jean et Sebastian Cabot.*

⁵³Rotz, *The Maps and Text of the Boke of Idrography,* 39 ff.

⁵⁴See note 41. Nicholas Barker, in his review of the Roxburghe Club reproduction of Rotz's atlas, edited by Helen Wallis, points out (as did Wallis) that Rotz drew his maps so that each leaf continues geographically where the previous one ends; but that this may not imply an exact following or size of the Dieppe or any other previous map. (*Times Literary Supplement,* May 6, 1983, p. 472.)

⁵⁵For details concerning the dispersal of Ralegh's maps, see R. A. Skelton, "Ralegh as a Geographer," *Virginia Magazine of History and Biography,* 71 (1963), 149, hereinafter cited as Skelton, "Ralegh as a Geographer." A list of some 500 books Ralegh had with him in the Tower, used in preparing his *Historie of the World,* has been identified by Dr. Walter Oakeshott, rector of Lincoln College, Oxford: "An unknown Raleigh MS: the working papers for 'The History of the World,' " *Times* (London), November 29, 1952; and Walter Oakeshott, *The Queen and the Poet* (London: Faber and Faber, 1960), 17-20. Most of this library was collected by Ralegh for use in writing his *Historie,* which never reached beyond 150 B.C.; but the list included Münster's *Cosmographia,* with its two Verrazzano-type maps showing a sea or bay bisecting North America, and three volumes on America, possibly Theodorus de Bry's first three volumes of the *Grands Voyages, Collectiones Peregrinationum in Indiam Occidentalem* (Frankfort: 25 parts, 1590-1634).

⁵⁶Skelton, "Ralegh as a Geographer," 130-149, especially pp. 133, 135, 140, and 143.

⁵⁷E. S. de Beer (ed.), *The Diary of John Evelyn* (Oxford, England: Clarendon Press, 6 volumes, 1955), IV, 215.

⁵⁸Skelton, "The Cartography of the Voyages," 158-159, 278.

⁵⁹Skelton, "The Cartography of the Voyages," 280.

⁶⁰Helen Wallis, "The Royal Map Collections of England," *Centro de Estudos de Cartografici Antiga* (Coimbra: Junta de Investigacões Cientificas do Ultramar, 1981), CXLI, 1-10, hereinafter cited as Wallis, "The Royal Map Collections."

⁶¹R. A. Skelton, "The Royal Map Collections of England," *Imago Mundi,* XIII (1956), 181-183, hereinafter cited as Skelton, "The Royal Map Collections of England," notes twenty-eight manuscript maps and forty printed maps in the Old Royal Library listed by the librarian, Thomas Ross, in 1660, none of which is in the British Museum.

⁶²See plate 8. Johne Rotz, *The Boke of Idrography, 1542.* London, British Library, Royal MS 20 E-IX. This work, edited by Dr. Helen Wallis and published as *The Maps and Text of the Boke of Idrography Presented by Johne Rotz to Henry VIII,* has been previously discussed; see note 41 and elsewhere. Ganong, *Crucial Maps of the Atlantic Coast of Canada,* 221-224, 292-295, analyses the northern part of the North American coast. This map (fols. 23v-24r) has been reproduced in color in Cumming, Skelton, and Quinn, *The Discovery of North America,* 108.

⁶³Skelton, "The Royal Map Collections of England," 135. The "Harleian" map did not go to the British Museum with the Harleian MSS. It was apparently at Whitehall, where the butler made off with it. Sir Joseph Banks acquired it and presented it to the museum in 1790. See *Catalogue*

of MS Maps in the British Museum (London, 1844), Volume I, under Add. MS. 5413. For bibliography, see Layng, *Sixteenth-Century Maps Relating to Canada,* 82-83, No. 241.

[64]See plate 12 (BM add. MS. 5413) and plate 23 (White). The "Harleian" map is reproduced in color in Cumming, Skelton, and Quinn, *The Discovery of North America,* 162-163.

[65]A tabulation of the names on the chart is given in Marcel Destombes, "Nautical Charts Attributed to Verrazzano (1524-1528)," *Imago Mundi,* XI (1954), 64-65, hereinafter cited as Destombes, "Nautical Charts Attributed to Verrazzano," with an illegible reproduction (p. 57).

[66]Destombes, "Nautical Charts Attributed to Verrazzano," 61-62. Since the general outline and nomenclature is derived from a padrón type, Destombes dates the Greenwich portolan 1528, with additions made about 1539-1540. But the basic coastline is of a padrón type found on no Spanish map before 1529. Whether Destombes's ascription of authorship to Gerolamo is correct or not, there is no reason to believe that the Greenwich portolan 1528 was made before the late 1530s.

[67]Ganong, *Crucial Maps of the Atlantic Coast of Canada,* 72.

[68]Amando Cortesão and Avelino Teixeira da Mota, *Portugaliae Monumenta Cartographica* (Lisbon: n.p., 6 volumes, 1960[-1962]), II, Homem 1558, hereinafter cited as Cortesão and Teixeira da Mota, *Potugaliae Monumenta Cartographica*; Wallis, "The Royal Map Collections," 484.

[69]Ganong, *Crucial Maps of the Atlantic Coast of Canada,* 74-76, 346.

[70]For a reproduction of the East Coast detail and a full description of the map, including the nomenclature of the southeastern coast, see Cumming, *The Southeast in Early Maps,* plate 16 and p. 115. For additional reproductions, see Quinn, *New American World,* II, plate 76, and Seymour J. Schwartz and Ralph E. Ehrenberg, *The Mapping of America* (New York: H. N. Abrams, 1980), plate 29.

[71]Paris, Bibl. Nat. S. H. Arch., No. 2. A 1551 world map by Sancho Gutiérrez, younger son of Diego, Sr., and also a cosmographer of the Casa, is probably a closer copy of the padrón general than either of the father's maps. See Henry Wagner, "A Map of Sancho Gutiérrez of 1551," *Imago Mundi,* VIII (1951), 47. The author's examination of the original in 1958 in the Östreichische Bibliothek (Vienna, Austria), No. 200-262, shows that the Carolina coast area is water damaged and illegible for reproduction.

[72]Ganong, *Crucial Maps of the Atlantic Coast of Canada,* 395-399.

[73]Harrissee, *The Discovery of North America,* 189-197; Lowery, *The Spanish Settlements,* II, 3-27.

[74]Clifford M. Lewis and Albert J. Loomie, *The Spanish Jesuit Mission in Virginia, 1570-1572* (Chapel Hill: University of North Carolina Press, 1953), 12, hereinafter cited as Lewis and Loomie, *The Spanish Jesuit Mission in Virginia.*

[75]Quinn, *New American World,* II, 406.

[76]Shea, "Ancient Florida," in Winsor, *Narrative and Critical History,* II, 260.

[77]Paul E. Hoffman, *Spain and the Roanoke Voyages* (Raleigh: America's Four Hundredth Anniversary Committee, North Carolina Department of Cultural Resources, 1987).

[78]Lewis and Loomie, *The Spanish Jesuit Mission in Virginia,* 14-15.

[79]Lowery, *The Spanish Settlements,* II, 28-32; William P. Cumming, "The Parreus Map (1562) of French Florida," *Imago Mundi,* XVII (1963), 29, hereinafter cited as Cumming, "The Parreus Map."

[80]Quinn, *New American World,* 294.

[81]All but a few small ships of Ribault's fleet were wrecked in a hurricane. Their crews surrendered to Menéndez, who wrote to Philip II, "I put Ribaut and all the rest of them to the knife" (Lowery, *The Spanish Settlements,* II, 200). Jacque Ribault, the son of Jean Ribault, escaped to France with two ships after retrieving some crewmen who had escaped from La Caroline, among them Laudonnière and the painter le Moyne. Menéndez later attempted to justify his ruthless extermination of the French on the grounds that he had neither provisions to feed nor prisons and guards to secure them (Quinn, *The Roanoke Voyages,* II, 471-472). He continued, however, mercilessly to hunt them down and exterminate all stragglers found.

⁸²Le Moyne's own narrative, published in Theodorus de Bry (comp.), *Collectio Peregrina-tionum . . . in Indiam Occidentalem,* Volume II [*America,* Part II]. *Brevis Narratio . . . Auctore Jacobo de Moyne* (Francofvrti ad Moenum, 1591), II, 6, hereinafter cited as de Bry, *Collectio Peregrinationum* (1591); English translation in Paul Hulton (ed.), *The Work of Jacques Le Moyne de Morgues, a Huguenot Artist in France, Florida, and England* (London: British Museum Publications, Ltd., 2 volumes, 1977), I, 119, hereinafter cited as Hulton, *The Work of Jacques Le Moyne.*

⁸³Cumming, "The Parreus Map," 27-40.

⁸⁴Hulton, *The Work of Jacques Le Moyne,* I, 119. Le Moyne's meeting with and relationship to White and Ralegh are examined in Volume I, pp. 10-11.

⁸⁵The best commentaries on the le Moyne map are in Hulton, *The Work of Jacques Le Moyne,* I, 45-54, and Cumming, *The Southeast in Early Maps,* 12-13, 124-125. The latter work also examines its influence on later maps: 126 (Jode); 129-131 (Mercator-Hondius 1606); 137-138 (Blaeu 1740); 149-150 (Montanus 1671), ff. William P. Cumming, "Geographical Misconceptions of the Southeast in the Cartography of the Seventeenth and Eighteenth Centuries," *Journal of Southern History,* IV (1938), 476-492, hereinafter cited as Cumming, "Geographical Misconceptions of the Southeast," includes a study of the cartographic influence of the le Moyne map with a list of later regional maps that derived from it.

⁸⁶Lowery, *The Spanish Settlements,* II, 314-336, gives a clear account of the Gourgues episode.

⁸⁷Lowery, *The Spanish Settlements,* II, 212-213; Lewis and Loomie, *The Spanish Jesuit Mission in Virginia,* 18-20, has an excellent analysis of this theory that includes a map (p. 19) with an attempted reconstruction of Menéndez's geography of North America; see also L. A. Vigneras, "A Spanish Discovery of North Carolina in 1566," *North Carolina Historical Review,* XLVI (October, 1969), 398-401, hereinafter cited as Vigneras, "A Spanish Discovery."

⁸⁸Quinn, *New American World,* II, 400, 416.

⁸⁹Lewis and Loomie, *The Spanish Jesuit Mission in Virginia,* 22 n. 50.

⁹⁰Quinn, *New American World,* II, 546.

⁹¹Quinn, *New American World,* II, 255.

⁹²Lewis and Loomie, *The Spanish Jesuit Mission in Virginia,* 13.

⁹³Quinn, *New American World,* II, 257.

⁹⁴The river that they entered and possessed was called Rio de San Bartolomé; this name appears on no later map or rutter along this coast.

⁹⁵Vigneras, "A Spanish Discovery," 398-407, includes, along with the text of the pertinent documents, an analysis of this 1566 expedition and of Captain Pedro de Coronas's taking possession of the Carolina coast. Lewis and Loomie, *The Spanish Jesuit Mission in Virginia,* had given an account largely from secondary sources that accused the friars and crew of concocting the entire story of the storms in order to avoid establishing the mission and outpost. Quinn, *New American World,* II, 550-555, prints the chief documents and suggests that the latitudes (37°30′ and 36°) were half a degree too high, making the first landfall at the Chesapeake and the second somewhat north of Cape Hatteras.

⁹⁶Lewis and Loomie, *The Spanish Jesuit Mission in Virginia,* 196. Lewis and Loomie's account of the Spanish Jesuit mission to the Chesapeake in 1570, with pertinent documents transcribed and translated, is the fullest and most careful study of the subject. The best surviving account of Gonzalez's 1588 journey is that in Father Luis Gerónimo de Oré's *Relacion,* which is probably based on original documents copied in Santa Elena while Gerónimo de Oré was there in 1614 or 1616; Lewis and Loomie, 185-189, has a reprint of the translation in Father Maynard J. Geiger, *The Martyrs of Florida* (New York: J. F. Wagner, 1937), 44-48. See also Paul E. Hoffman, "New Light on Vicente Gonzalez's 1588 Voyage in Search of Raleigh's English Colonies," *North Carolina Historical Review,* LXIII (April, 1986), 199-223.

⁹⁷Lewis and Loomie, *The Spanish Jesuit Mission in Virginia,* 263, plate XI.

⁹⁸Lewis and Loomie, *The Spanish Jesuit Mission in Virginia,* 264, 265.

[99]Numerous exceptions to this general rule can be cited: exiles and expatriates such as the Portuguese Homems took unauthorized information with them or gave details of voyages, discoveries, or settlements.

[100]For Spanish and especially French and Portuguese cartographic additions and information on the Canadian-Newfoundland coasts, as well as valuable critical comments on many cartographers discussed here, see Ganong, *Crucial Maps of the Atlantic Coast of Canada*. Hoffman, *Cabot to Cartier*, also includes numerous contributions concerning the Northeast, although the author's interest is primarily ethnological.

[101]Reprod., plate 16 (detail); names on Southeast coast, in appendix; names on New England coast, Cumming, "The Colonial Charting of the Massachusetts Coast," Appendix B. Endorsed: "Gulf and Coast of New Spain from the papers of Alonso de Santa Cruz that they took from Seville." Presently in the Archivo Historico Nacional, Madrid. Dated (incorrectly) 1536 and reproduced in *Mapas Españoles de America* (Madrid, 1951), Plate IX. See descriptive analysis in Cumming, *The Southeast in Early Maps*, 114-115; reprod. in Cumming, Skelton, and Quinn, *The Discovery of North America*, 172.

[102]Lowery, *The Spanish Settlements*, II, 245 ff.

[103]Cumming, Skelton, and Quinn, *The Discovery of North America*, 70: from a copy of Santa Cruz's *Islario General*, ca. 1545, in the Archivo Historico Nacional, Madrid.

[104]For the most active area of Spanish coastal occupation from the cape to Santa Elena (Port Royal, South Carolina), a few maps have survived. See Woodbury Lowery, *The Lowery Collection: A Descriptive List of Maps of the Spanish Possessions within the Present Limits of the United States, 1502-1820*, edited with notes by Philip Lee Phillips (Washington, D.C.: Government Printing Office, 1912), hereinafter cited as Lowery, *The Lowery Collection*, and a reproduction of "Mapa de la Florida y Laguna de Maymi," ca. 1595-1600, a manuscript map in the Archives of the Indies, Seville, in Lowery, *The Spanish Settlements*, II, opposite p. 286.

[105]Quinn, *The Roanoke Voyages*, I, 201.

[106]Ganong, *Crucial Maps of the Atlantic Coast of Canada*, 450, first suggested Fernándes as a possible compiler, but Ganong was confused as to the identity of "Fernando Simon" on the Fernándes-Dee map; he likewise confusedly attributed an Italian map by Milõ to Fernándes (Figure 106; corrected in the second [1964] edition on p. 495).

[107]British Library, Cotton MS. Augustus I.i.i. Reprods.: Hakluyt, *The Principal Navigations*, VIII, folded at end, entire map; E. G. R. Taylor (ed.), *The Writings and Correspondence of the Two Richard Hakluyts* (London: Hakluyt Society, second series, LXXXVI-LXXXVII, 2 volumes, 1935), II, end map of North America; Humphrey Gilbert, *The Voyages and Colonizing Activities of Sir Humphrey Gilbert*, edited by David Beers Quinn (London: Hakluyt Society, second series, LXXXIII, LXXXIV, 1940), II, end pocket, enlarged coastal section north of Cape Trafalgar, hereinafter cited as Gilbert, *Voyages and Colonizing Activities*.

[108]E. G. R. Taylor, *Tudor Geography, 1485-1583* (London: Methuen and Co., 1930), hereinafter cited as Taylor, *Tudor Geography*; and Taylor, *Late Tudor and Early Stuart Geography, 1583-1650* (London: Methuen and Co., 1934), passim. These studies have done much to reestablish Dee's reputation and show his contributions.

[109]Reprod.: Cumming, Skelton, and Quinn, *The Discovery of North America*, 126-127, plate 140. A similar promontory is found on the Fernándes-Dee chart of 1580 and on John White's general map of the Southeast. Diogo Homem was related to Lopo and Andre; the cartographic work of the three Homems often exhibits a common Portuguese basis.

[110]Ganong, *Crucial Maps of the Atlantic Coast of Canada*, 450.

[111]Skelton, "Ralegh as a Geographer," 133. On the back of this map Dr. John Dee presented a claim for "the great part of the seacoast of the Atlantic (otherwise called America) and all the Isles next Northerly as due and appropriate unto our soveraigne Elizabeth," citing twelve reasons. Among these were:

"1. Lord Madoc, sonne to Gwynedd, led a colony and inhabited Florida circa an. 1170.
2. Robert Thorne his father and Mr. Eliot of Bristol, 1494.

3. Brandan 560 (cf. S. Brandani an island on the map SE of Cape Breton: Sable I).

4. Sebastian Cabot sent by Henry VII. 1497."

The claims were not so forceful as to convince the Spanish king or his ministers; nor did Lord Burghley react favorably after reading it. For October 3, 1580, as the date of Dee's presentation of the map to Queen Elizabeth, see Taylor, *Tudor Geography*, 122, 134-135, 188.

[112]British Library. MS Cott. Rol., XIII. 48. Vellum; size: 32′ x 47¼. Full repr.: Cortesão and Texeire da Mota, *Portugaliae Monumenta Cartographica*, II, 129-131, plate 240. The toponymy obviously indicates the area of Dr. Dee's current interest. It has been suggested that Fernándes had drawn his own copy, but no significant changes are noted from the usual Portuguese type that would indicate any special knowledge incorporated by Fernándes. It is more likely, given what is known of Fernándes's career, that he "liberated" the chart while looting some vessel such as the great Portuguese caravel that he seized off the Azores in 1574. (David B. Quinn, *England and the Discovery of America, 1481-1620* [New York: Alfred A. Knopf, 1974], 148-149, hereinafter cited as Quinn, *England and the Discovery of America*.)

[113]Quinn, *The Roanoke Voyages*, I, 94.

[114]"S. John" is also found on the Dee chart of 1580, below C. de las Arenas.

[115]Fernándes's own signature appears on a letter to Martin Frobisher requesting equipment in 1581 for the *Galleon* for a voyage that Fenton, not Frobisher, later commanded with Fernándes as pilot (E. G. R. Taylor, *The Troublesome Voyage of Captain Edward Fenton, 1582-1583* [Cambridge: Cambridge University Press, 1959], 22, hereinafter cited as Taylor, *The Troublesome Voyage*).

[116]Quinn, *England and the Discovery of America*, 243-283, has an excellent summary of Fernándes's life and includes new documentary information from Spanish and English sources.

[117]Vigneras, "A Spanish Discovery," 408-412, gives a list of the twenty men appointed by Menéndez to go on the expedition of *La Trinidad* in 1566.

[118]Taylor, *The Troublesome Voyage*, 38.

[119]Gilbert, *Voyages and Colonizing Activities*, I, 187, quoting *Calendar of State Papers, Spanish, 1568-1579* (1894), No. 496.

[120]Richard Madox, *An Elizabethan in 1582: The Diary of Richard Madox, Fellow of All Souls*, edited by Elizabeth S. Donno (London: Hakluyt Society, second series, No. 147, 1976), 33, 44.

[121]Quinn, *The Roanoke Voyages*, I, 94. Estimates of distances, especially while sailing along a coast, should always be regarded with caution; but the English mile of the sixteenth century, unlike the variable league, closely approximated the modern English mile. (Ronald Edward Zupko, *A Dictionary of Weights and Measures: From Anglo-Saxon Times to the Nineteenth Century* [Madison: University of Wisconsin Press, 1968], 106.)

[122]Quinn, *The Roanoke Voyages*, I, 94.

[123]No maps from the 1584 voyage have survived, although it is possible that Ralegh received one. Barlowe indicated in his report that members of his expedition had found a place for a pearl fishery: "In what places of the Countrey the pearle grew, which nowe your Worshippe doth very well understand." (Quinn, *The Roanoke Voyages*, I, 105.)

[124]Lane evidently meant that none of the inlets could accommodate large ships.

[125]Quinn, *The Roanoke Voyages*, I, 201-202.

[126]Quinn, *The Roanoke Voyages*, I, 106-107.

[127]For support of Trinity harbor or a nearby inlet, see Talcott Williams, "The Surroundings and Site of Raleigh's Colony," *Annual Report of the American Historical Association, 1885* (Washington, D.C.: United States Government Printing Office, 1896), and C. W. Porter III, "Fort Raleigh," *North Carolina Historical Review*, XX (January, 1943), 25. Porter is inclined to favor an inlet near Nags Head (which he incorrectly identified with Kenricks Mounts below Port Ferdinando because of Barlowe's reference to grapes). He notes that William Byrd also suggested a place near Collington Island as the site of Barlowe's landing. Conway Whittle Sams, *The Conquest of Virginia: The First Attempt* (New

York: G. P. Putnam's Sons, 1939), 81-83, with little apparent justification, places Barlowe's entry at Ocracoke: "Captain Barlowe says . . . they saw before them 'another mightie sea,' 'another great sea.' This of course is Pamlico Sound. . . . At Ocracoke one would be some twenty-eight miles from the mainland. We therefore conclude that Ocracoke Inlet was the one through which the vessels passed into Pamlico Sound."

[128]David Stick, *The Outer Banks of North Carolina* (Chapel Hill: University of North Carolina Press, 1958), 14, hereinafter cited as Stick, *The Outer Banks.* The mouth of Jeanquite or Jean Creek is about two miles north of Wright Bridge and in the approximate location of Trinity harbor.

[129]Quinn, *The Roanoke Voyages,* I, 95 n. 1, 106, 201, 415; Morison, *The European Discovery of America,* 620, 621, 625.

[130]Quinn, *England and the Discovery of America,* 255. Still another deposition (presently in the Archives of the Indies, Seville) was made by a member of the 1585 voyage, a castaway on Jamaica (*England and the Discovery of America,* 256-257). It is not clear what voyages this Englishman refers to. He says that Fernándes had intended (in 1578?) to explore Florida but did not reach it, that when the deponent reached a bay with islands the Indians on one promontory ate thirty-eight Englishmen but that the Indians on the other promontory were gentler, that there was much gold and silver in the region, and that the Indians gave the Portuguese (Fernándes) four pounds of gold and a hundred of silver. This is evidently a confused and garbled story, probably combining hearsay from different voyages. It may indicate that Fernándes did *not* explore "Florida" before 1584, that some hostilities developed with the Indians (although not a massacre of thirty-eight English) after the Englishmen landed (in 1584?)—hostilities that were suppressed in the published accounts—and that Chesapeake Bay may have been visited by Fernándes's ship after it left Roanoke on one of the voyages.

[131]Quinn, *The Roanoke Voyages,* I, 202.

[132]In Morison, *The European Discovery of America,* 640, the author is vigorous in denunciation: "Ferdinando committed the pilot's unforgivable sin. While conning *Tiger* through the inlet he ran her aground. . . . The most charitable thing we can say about Simon is that he never learned his business." Morison suggests (p. 631) that Fernándes had been suborned by the Spanish secret service to wreck the expedition; this is a most unlikely surmise, since for the previous decade Fernándes had been a major scourge of Spanish shipping and was a promoter of the Ralegh colony.

[133]Quinn, *The Roanoke Voyages,* I, 201. This sentence is crossed out in the manuscript. Lane's deletion may indicate that he had, seven weeks after arriving, decided that Port Ferdinando was the main point of entry or had heard of another bay farther north (Chesapeake).

[134]The location of Wococon (Ococon) at or near the present Ocracoke Inlet, rather than farther south on narrow and broken Portsmouth Island (compare with Quinn, *The Roanoke Voyages,* II, 867), is based on the following reasoning. The size and shape of land on both sides of the entry on the sketch-map, the White manuscript, and the White-de Bry engraving is wide, with an inward curve on the northern island that makes a small bay. This is similar to Silver Lake, the harbor of the present village of Ocracoke, and in the same location on the island. The directional relation of Wococon to the mainland tributaries (the Neuse and Pamlico rivers) conforms to the present location of Ocracoke Inlet. The size and shape of the barrier islands south of Cape Hatteras change less than those to the north, which are subject to greater wind and wave force. (Quinn, *The Roanoke Voyages,* II, 867.)

[135]Quinn, *The Roanoke Voyages,* I, 259. Evidently this island is the one placed by White on his large-scale map; no such island exists. Quinn suggests that Menatonon, the Indian chief, was referring to Kecoughtan (Old Point Comfort) as the island. (*The Roanoke Voyages,* I, 259-260 n. 5.)

[136]Quinn, *The Roanoke Voyages,* I, 251-258, 261, 285-286. Lane's conception and account of Chesapeake Bay and the expedition there are confused. In the report made upon his return to England, he wrote: "I woulde haue sent a small Barke with two Pinnesses about by Sea to the Northwarde to haue found out the Bay he spake of," while with a hundred men Lane would have gone from the head of Chowan River "by foote all the voyage ouer land" (*The Roanoke Voyages,* I, 261-262).

74

When, with how many troops, and by what route the expedition went to Chesapeake, Lane fails to note; but the "Colonel of the Chesepians" (Philip Amadas?) was back at Roanoke by June 1, for he shot Wingina (Pemisan), the Indian chief, during a fight between the Indians and the colonists. (*The Roanoke Voyages*, I, 286-287.)

[137]Quinn, *The Roanoke Voyages*, II, 523. Lane's pessimistic view in his final report was that, in spite of enjoying "the most sweete, and healthfullest climate . . . in the world," inducement to settle would require "the discouery of a good mine . . . or a passage to the Southsea, or someway to it, and nothing els. . . ." (*The Roanoke Voyages*, I, 273.)

[138]David B. Quinn, "Turks, Moors, Blacks, and Others in Drake's West Indian Voyage," *Terrae Incognitae*, XIV (1982), 97-104.

Baptista Boazio's general map of "The famous West Indian voyadge" of Sir Francis Drake's destructive and lucrative raids upon Spanish ports shows "Virginia." Virginia, however, is placed nearer Cape Cod than the Outer Banks. Color reprod. in Cumming, Skelton, and Quinn, *The Discovery of North America*, 186.

Boazio was page to Lieutenant General Christopher Carleill, captain of the *Tiger* during Drake's West Indian expedition; White returned aboard the *Tiger* from Virginia in 1586. Thus Boazio must have seen White's drawings, several of which he copied for his maps, including the Sea Connye (trigger fish) at the bottom left of his general map. See Mary F. Keeler, *Sir Francis Drake's West Indian Voyage* (London: Hakluyt Society, 1981), 314-319, especially 319 n. 4. Unfortunately, Boazio did not use White's maps for his erroneous east coast, possibly to conceal the actual location of the colony.

Another contemporary map, thought to be a duplicate of the one Drake presented to Queen Elizabeth, which like Boazio's has only "Virginia" in back of an ill-defined coast with misplaced indication of Drake's landing there, is the so-called Drake-Mellon map in the collection of Paul Mellon of Upperville, Virginia (reproduced in *Sir Francis Drake*, the catalog of the Drake Exhibit, published by the British Museum in 1977, p. 80, No. 75).

[139]Quinn, *The Roanoke Voyages*, II, 523.

[140]Quinn, *The Roanoke Voyages*, II, 523.

[141]White's complaints against Fernándes are in the report taken from his journal, which was published in Hakluyt's *principall navigations voiages and discoueries of the English nation* (1589). See Quinn, *The Roanoke Voyages*, II, 518, 520, 521, 522, 523, 525.

[142]Quinn, *The Roanoke Voyages*, II, 515 ff. The still-lively controversy concerning the fate of the "Lost Colony" is of no concern here. William Strachey, who published a history of the Jamestown colony in 1612, wrote that, about 1606, Powhatan had "miserably Slaughtered . . . the men, women, and Children of the first plantation at Roanoke." William Strachey, gent., *The Historie of Travell into Virginia Britanica (1612)*, edited by Louis B. Wright and Virginia Freund (London: Hakluyt Society, second series, CIII, 1953), 91, hereinafter cited as Strachey, *The Historie of Travell into Virginia Britanica*.

[143]The Spanish royal cosmographer Alonso de Chaves placed the Baya de Santa Maria at 36½° to 37° in his *derrotero* of 1539 (see appendix).

[144]Quinn, *England and the Discovery of America*, 255.

[145]Irene A. Wright (ed.), *Further English Voyages to Spanish America, 1583-1594* (London: Hakluyt Society, second series, XCIX, 1951), 175-176, hereinafter cited as Wright, *Further English Voyages*; Quinn, *The Roanoke Voyages*, I, 80-81. Quinn suggests tentatively that Edward's testimony may refer to a hostile Indian reception (not cannibalistic) in Chesapeake Bay if Amadas's flagship with Fernandes as pilot went north after leaving Port Ferdinando (Quinn, *England and the Discovery of America*, 256-257).

[146]Strachey, *The Historie of Travell into Virginia Britanica*, 131. Professor Quinn comments: "The evidence is suggestive rather than conclusive, but it remains just possible that Fernández's influence on English ventures in these parts of North America lasted long after his disappearance or death." (Quinn, *England and the Discovery of America*, 161.)

[147]Quinn, *The Roanoke Voyages*, I, 94.

[148]Quinn, *The Roanoke Voyages,* I, 293. Quinn suggests that the loss of goods may have resulted from their being tipped overboard by sailors, impatient of any delays (Quinn, *England and the Discovery of America,* 301).

[149]Quinn, *The Roanoke Voyages,* I, 94.

[150]There are six sounds north of Cape Fear:

Currituck: a narrow, shallow, multi-islanded sound stretching between the barrier islands and the mainland and extending north into Virginia from Albemarle Sound. Both Currituck and Trinity inlets were in use by the colonists; both are presently closed.

Albemarle: a roughly rectangular sound west from Kitty Hawk and Colleton Island that extends more than fifty miles to the mouths of the Roanoke and Chowan rivers. The colonists called it Chowan and Occam (eastern part).

Croatan: between Roanoke Island and the mainland; marshy and so dotted with islands that one could at times walk from the southern end of the island to the mainland with the aid of a fence rail to cross one gut or slough. Gary Dunbar, *Historical Geography of the North Carolina Banks* (Baton Rouge: Louisiana State University Press, 1958), 137 n. 7, hereinafter cited as Dunbar, *Historical Geography of the North Carolina Banks.* As Croatan Sound received more of the flow of water from Albemarle Sound, it deepened.

Roanoke: between the Outer Banks and Roanoke Island; narrow, islanded, and shallow.

Pamlico: south of Roanoke Island and west of Ocracoke Inlet to the estuaries of the Pamlico-Tar and Neuse rivers. Largest of the sounds: the mainland is fifteen to twenty-six miles from the barrier islands; the sound forms a wide expanse with a few islands close to the land.

Core: a narrow, shallow sound from Pamlico Sound to Cape Fear, situated between low-lying Core Banks and the mainland. Averaging between two and three miles in width, it has many islands.

[151]Lewis and Loomie, *The Spanish Jesuit Mission in Virginia,* 12-13. Two or more wrecks were recorded in some years.

[152]Quinn, *The Roanoke Voyages,* I, 302.

[153]*United States Coast Pilot* (1980), 4. Cape Henry to Cape Lookout, Chart 11555.

[154]The Gulf Stream joins the Antilles Current at the Straits of Florida and generally follows the 100-fathom contour to Cape Hatteras, where its 50-mile width broadens and slows from an average speed of 2.5 nautical miles per hour (4 nautical miles per hour along the Florida coast) to 5 nautical miles per day, according to the *United States Coast Pilot* (1980).

[155]It is thought that during such periods of submergence the Gulf Stream may have flowed undiverted until it reached New England.

[156]Vincent Bellis, Michael P. O'Connor, and Stanley R. Riggs, *Estuarine Shoreline in the Albemarle-Pamlico Region of North Carolina* (Raleigh: North Carolina State University, 1975), 15-26. This pamphlet gives a clear view, with diagram, of the various changes in the North Carolina coastline resulting from erosion.

[157]Armin K. Lobeck, *Geomorphology: An Introduction to the Study of Landscapes* (New York: McGraw-Hill Book Company, 1939), 345.

[158]Robert Dolan and Kenton Bosserman, "Shoreline Erosion and the Lost Colony," *Annals of the Association of American Geographers,* 62 (1972), 424-426, hereinafter cited as Dolan and Bosserman, "Shoreline Erosion," estimates maximum erosion along most of the Outer Banks between 1585 and 1972 at about a quarter mile.

[159]Quinn, *The Roanoke Voyages,* II, 617. The cape is named "C. Kenrick" on the Velasco map of 1611 (*The Roanoke Voyages,* II, Figure 10, opposite p. 851, and p. 864, No. 53) but is misnamed "Cape Hatterass" on the Comberford (1657) chart. Ogilby's (ca. 1672) and Gascoyne's (1682) maps show only Cape Hatteras; the erosion of the second cape may have occurred shortly before the last quarter of the seventeenth century.

[160]Captain James Wimble charted the shoals and named them after himself on his *Chart . . . of North Carolina.* William P. Cumming, "Wimble's Maps and the Colonial Cartography of the North

Carolina Coast," *North Carolina Historical Review,* XLVI (April, 1969), plate 2, hereinafter cited as Cumming, "Wimble's Maps." Wimble shows an inlet at this place.

    161Stick, *The Outer Banks,* 6.

    162G. L. Giese, H. B. Wilder, and G. G. Parker, Jr., *Hydrology of Major Estuaries and Sounds of North Carolina* (Raleigh: United States Geological Survey: Water Resources Investigation, 79-4-6, 1979), 75, table 3.1: 31,700 cubic feet per second.

    163A historical tabulation of the Outer Banks is beyond the scope of this work. Several historical lists, with notations, have been made; existing long before them were articles on different inlets. In April, 1806, Secretary of the Treasury Albert Gallatin appointed three commissioners—Thomas Coles, Jonathan Price, and William Tatham—to survey the coast of North Carolina from Cape Hatteras to Cape Fear. Tatham, a brilliant but erratic English surveyor, lawyer, and scientist, showed contempt for Coles and Price, the two North Carolina commissioners; they reciprocated and surveyed independently. Most of the notes and surveying instruments of all three were lost with the sinking of their two cutters in a hurricane on the night of September 28, 1806. Coles and Price made "A Chart of the Coast of North Carolina between Cape Hatteras and Cape Fear," with notes (presently in the National Archives). Tatham made a 54-page report with details listing the inlets together with a map (presently in the Library of Congress). See G. Melvin Herndon, "The 1806 Survey of the North Carolina Coast, Cape Hatteras to Cape Fear," *North Carolina Historical Review,* XLIX (July, 1972), 242-253. For additional details and references, see Cumming, "Wimble's Maps," 57, n. 3. A list of thirty inlets from Virginia to the Cape Fear River, with map and dates on which they appear, is in *Additional Inlets,* published (1923) by the North Carolina Fisheries Commission; this early study has much useful reminiscent detail furnished by old Banks natives and not preserved elsewhere. For inlets recorded in the 1585-1590 maps and reports, the standard is still Quinn, *The Roanoke Voyages,* II, 852-872.

    Gary Dunbar's *Historical Geography of the North Carolina Banks* is the standard in its field; numerous comments and references on various inlets are scattered through its notes, and its appendix A ("Inlet Changes in Historic Time" [pp. 215-217]) examines eleven inlets with an accompanying map. David Stick's *Outer Banks* has a table of twenty-four inlets, with various names given to the same inlet, locations, and dates of openings and closings. The fullest study of the inlets is John Joseph Fisher, "Geomorphic Expression of Former Inlets along the Outer Banks of North Carolina" (unpublished master's thesis, University of North Carolina at Chapel Hill, 1962), hereinafter cited as Fisher, "Geomorphic Expression of Former Inlets." Several typescript copies of this work are in UNC-CH libraries. The work's Appendix 2, pp. 83-112, contains charts and an annotated historical account of each inlet.

    164Some modern cartographic attempts to show the location of inlets in 1585 have erroneously placed Wococon at Whalebone Inlet (a cut in the narrow southern part of Portsmouth Island that opens occasionally) and have shown no inlet at Ocracoke. This error may be connected with or based upon a mistaken reading in 1905 of the White MS B by the respected geologist Collier Cobb, who wrote that an "Onoaconan" inlet was probably present Ocracoke Inlet and that a "Wococon," found on other maps, was probably Whalebone Inlet (Collier Cobb, "Some Changes in the North Carolina Coast Since 1585," *North Carolina Booklet,* IV [January, 1905], 5-6). But as John Joseph Fisher's "Geomorphic Expression of Former Inlets" points out, "The map [White, La Virgenia Pars (plate 23)] Cobb must have originally consulted shows that he mistook an island for the letter 'O' and misread a 'W' as a 'no,' with the result that the inlet 'Onoaconan' was created from the name 'Wococon.'" That the nonexistent "Onoaconan" and Wococon (White-de Bry 1590) both refer to present Ocracoke Inlet is clear because all three White maps place Wococon in the same directional relationship to the mainland as modern Ocracoke Inlet; and both Portsmouth and Ocracoke islands are wide at the inlet, with a curved hook at the end of the Ocracoke Island corresponding to Silver Lake, the harbor of the modern town of Ocracoke. Whalebone Inlet could never have had such features.

    165William P. Cumming, "The Earliest Permanent Settlement in Carolina: Nathaniel Batts and the Comberford Map," *American Historical Review,* 45 (October, 1939), 89.

77

[166]Dolan and Bosserman, "Shoreline Erosion," 424-426 and Figure 1.

[167]Many writers have discussed various facets of the colony's difficulties and ultimate failure. The best general analysis of the problems relating to this colonial effort is probably presented in Quinn, *England and the Discovery of America,* chapter 2: "The Failure of Ralegh's American Colonies."

[168]Quinn, *The Roanoke Voyages,* I, 323.

[169]Quinn, *The Roanoke Voyages,* I, 158.

[170]Quinn, *The Roanoke Voyages,* I, 135. The manuscript (MS D/DRh, MI) is in the Essex County Record Office, County Hall, Chelmsford, England. Quinn has edited the text with careful notes: *The Roanoke Voyages,* I, 130-139. He identifies the author as Roger Williams, whose "A briefe discourse of warre" (1590) has very similar passages (Quinn, *New American World,* III, 274).

[171]The documents relating to this intended voyage are printed in Humphrey Gilbert, *The Voyages and Colonizing Activities of Sir Humphrey Gilbert,* edited by David Beers Quinn (London: Hakluyt Society, second series, II, 1940), 341-364.

[172]Quinn, *New American World,* III, 239-245. The original instructions are lost; they were copied by Sir Richard Hoby, who knew Sir Humphrey Gilbert, in his commonplace book (presently in the British Library, Add. MS. 38823 ff. 1-8). Except for instruments needed and subjects to be surveyed, methods are absent; evidently Bavin's competence as a surveyor is assumed.

[173]Quinn, *New American World,* III, 64-69. The original is in the British Library, Sloane MS 1447; it follows another set of "Inducements" for a voyage "to that parte of America which lythe betweene 34. and 36. degree." The first tract is clearly directed to the Roanoke voyage and written in early 1584. Although the title of the second "Inducements" appears to be directed toward a New England latitude, the text shows no implicit latitudinal emphasis; Quinn thinks it is "clearly concerned with the planning stages of the 1585 Virginia voyage" (*New American World,* III, 61).

[174]William P. Cumming, "The Identity of John White Governor of Roanoke and John White the Artist," *North Carolina Historical Review,* XV (July, 1938), 197-203, in which various aspects of White's identity are examined; Quinn, *The Roanoke Voyages,* passim; David Beers Quinn, *North America from the Earliest Discoveries to First Settlements: The Norse Voyages to 1612* (New York: Harper and Row, 1977), 325, hereinafter cited as Quinn, *North America from the Earliest Discoveries.*

[175]*Quinn, The Roanoke Voyages,* II, 712, 715, 715 n. 4. White's account of the 1590 voyage, published in Hakluyt's *Principal Navigations* (1600), 288 (reprinted in *The Roanoke Voyages,* II, 598), also begins "The fift voyage of Master John White into . . . Virginia in the Yeere 1590." It is doubtful that White would have repeated such a statement in a personal letter if it were not true, or that Hakluyt, a director of the expeditions, would have published it. In 1602 a John White, who could have been the governor making another attempt to search for the colonists, went aboard a privateer, the *Susan Parnell,* to the West Indies. He transferred to a Spanish prize vessel; but if he did so with the hope of returning to Virginia to search for the "Lost Colony," no record survives. Paul Hulton and D. B. Quinn, *The American Drawings of John White, 1577-1590* (London and Chapel Hill: University of North Carolina Press, 2 volumes, 1964), 22-23 (R. A. Skelton is the author of the cartographic essay "John White's Contribution to Cartography," pp. 52-57), hereinafter cited as Hulton and Quinn, *The American Drawings of John White*; Quinn, *England and the Discovery of America,* 446-447.

[176]Quinn, *The Roanoke Voyages,* I, 42-47, discusses several contemporary John Whites, such as the mayor of Plymouth in 1583-1584, who could have been a relation of Martin White, an agent of Ralegh in that port of departure for the 1584 voyage of discovery; and a John White commissioned in 1590 to survey and view the queen's storehouses in Plymouth, whence John White the governor set sail in 1587. John White the painter and governor was apparently a citizen of London and was certainly married, with at least one daughter, Eleanor. Further research may well uncover new information.

[177]British Museum, P. & D. 1906-5-9-1 (76 fols., including map). Although White's name is not given on the small title sheet or on any of the drawings, there is no question that they are all by him. Most of the drawings are, as the title states, "pictures of sondry things collected . . . in the voyage . . . for the discovery of La Virginea . . . 1585"; included in the portfolio, however, are draw-

ings of a number of figures having no relation to the Roanoke voyages (e.g., Eskimos and Turks). Quinn, *The Roanoke Voyages,* I, 392, shows that this set of drawings is not the original but is instead one of at least eight sets made by White of which there is some record. Quinn's *Roanoke Voyages,* I, 390-464, includes an annotated list of the White portfolio drawings together with other White original drawings, chiefly of North American subjects. The definitive work on the drawings is Hulton and Quinn's *The American Drawings of John White.* Volume I of this work contains critical and bibliographical information, and Volume II reproduces the originals handsomely in color facsimile. Quinn suggests that the set in the British Museum Department of Prints and Drawings was made by White between 1586 and 1590 for Ralegh, Grenville, or another important sponsor of the colony. *The American Drawings of John White,* I, 65-155, lists a total of 139 items by or attributed to White that have been found in various collections in the British Museum or elsewhere; those that are not duplicates of the drawings in the White portfolio throw no additional light on White's life or the Roanoke voyages.

[178]Hulton and Quinn, *The American Drawings of John White,* I, 13-14; Quinn, *North America from the Earliest Discoveries,* 324. If White went as painter in 1577, he must have made numerous drawings that have since disappeared, as have the original paintings of Virginia by White and of Florida by le Moyne. Ralegh may have chosen White in 1584 (or 1585) because he had seen or heard of his paintings of Eskimo life and characters.

[179]Hulton and Quinn, *The American Drawings of John White,* I, 76.

[180]Hulton, *The Work of Jacques Le Moyne,* I, 10, 12 n. 35.

[181]Hulton, *The Work of Jacques Le Moyne,* I, 11.

[182]St. Mary's Hall was affiliated with Ralegh's College of Oriel.

[183]Quinn, *The Roanoke Voyages,* I, 36-37, 37 n. 1. The remarkable achievements of Harriot, who, in Hakluyt's phrase, linked "theory with practice, not without incredible results," are examined in John W. Shirley (ed.), *Thomas Harriot, Renaissance Scientist* (Oxford: Oxford University Press, 1974), and clearly summarized in Charles C. Gillespie (ed.), *Dictionary of Scientific Biography* (New York: Charles Scribner's Sons, 16 volumes, 1970), VI, 124-129, hereinafter cited as *Dictionary of Scientific Biography.*

[184]Quinn, *The Roanoke Voyages,* I, 318-385, gives the text of Harriot's *Report* with extensive notes; pages 403-464 passim of the same volume has Harriot's comments on the engravings.

[185]Quinn, *The Roanoke Voyages,* I, 414 n. 5 and 430; instructions to Bavin in Quinn, *New American World,* III, 239-245.

[186]Harriot was also committed to the Tower with Northumberland for alleged complicity in the Gunpowder Plot (discovered on November 5, 1605) but was soon released for lack of evidence. Northumberland remained a prisoner until pardoned in 1622. Ralegh had been committed to the Tower on July 19, 1603, for conspiracies against King James I (which were never proved), although he was not beheaded until 1618.

[187]"Harriot," in *Dictionary of Scientific Biography,* VI, 124-129.

[188]The sketch could have been made by Amadas or by Cavendish, who possibly was experienced as a cartographer. (Hulton and Quinn, *The American Drawings of John White,* I, 54 n. 2.)

[189]Hulton and Quinn, *The American Drawings of John White,* I, 54.

[190]London, British Museum, P. & D., 1906-5-9-1 (2). Quinn, *The Roanoke Voyages,* I, 460-461, gives suggestions of sources and a list of all names on the map; Cumming, *The Southeast in Early Maps,* 15-17, 70, 118-121, provides commentaries on all the White maps and gives bibliographical descriptions and references; Hulton and Quinn, *The American Drawings of John White,* I, 52-57, examines the cartographic problems.

White's original drawings are in a set prefixed by the title "The pictures of sondry things collected and counterfeited according to the truth in the voyage made by [for?] S.[r] Walter Raleigh knight for the discovery of La Virginea. In the 27th yeare of the most happie reigne of our Soueraigne lady Queene Elizabeth and in the yeare of o.[r] Lorde God. 1585."

The earlier history of the drawings is unknown; the set was bought from Payne, the London bookseller, by the first earl of Charlemont in 1788 and sold by the third earl in 1865 through Sotheby to Henry Stevens, who in turn sold it to the British Museum in 1866. The Sotheby warehouse caught fire; the edges of the drawings were singed but saved by hosing; the soaked pages were put under pressure for three weeks. The large map was folded and compressed with no protective sheet between; this accounts for the strange duplication of a ship, a dolphin, a whale, etc., on the land, clearly visible in this reproduction.

A collection of drawings by White is in the British Library in the Sloane Collection; it contains no maps. (Quinn, *The Roanoke Voyages,* II, 914-915.)

[191]Hulton and Quinn, *The American Drawings of John White,* I, 56.

[192]Quinn, *The Roanoke Voyages,* I, 188-189. It is interesting, though perhaps not significant, that White drew ships—or a ship—following this route off Port Royal, beyond C. de S. Helene, Wococon, and finally at Port Ferdinando, where some of the men disembarked.

[193]The le Moyne map engraved by de Bry in 1591 has the southern shores of a large body of water at the top, but no strait connecting it to the Atlantic Ocean.

[194]Destombes, "Nautical Charts Attributed to Verrazzano," 60-62, with reproduction of the "Verrazzano" chart.

Two detailed photographs of the original "Verrazzano" chart are in the Cumming Map Collection in the Davidson College Library, Davidson, North Carolina.

[195]The names Jean Ribault gave to the islands and rivers along the coast have caused confusion and contradictory identifications for centuries. A Spanish spy's tracing of an original explorer's map made by Nicholas Barré, Ribault's pilot, has been found in the Museo Naval, Madrid. It is so detailed that definite identifications have been made. Cumming, "The Parreus Map," 27-40.

[196]The fullest and best analysis of the le Moyne-de Bry map of 1590 are Lowery, *The Spanish Settlements,* II, 410-417 (Appendix J), and that of Skelton in Hulton, *The Work of Jacques Le Moyne,* I, 45-54. Both Lowery and Skelton, it should be mentioned, noted the possibility that the many divergences between de Bry's and White's portrayals of French Florida may be explained by White's having used a map drawn by some other survivor of the French expedition, such as Laudonnière; but Skelton (Hulton, *The Work of Jacques Le Moyne,* 46) refers to the present writer's postulation (Cumming, "The Parreus Map," 39-40) of an earlier, different version by le Moyne used by White and a later one used by de Bry (and his imitators), which incorporated changes and additions from oral accounts or from printed narratives such as those of Laudonnière (1586).

[197]This is the present Caloosahatchee River, which is still the only natural outlet of Okeechobee except for the slow seepage through the Everglades. The representation of Okeechobee, except for these two maps and the later derivatives of the le Moyne-de Bry engraving, was unknown for more than one hundred years except for the possibility that a lake, "Maymi," in upper central Florida, on a Spanish map of about 1595, is intended for Okeechobee. It may be Lake George. See Cumming, *The Southeast in Early Maps,* 126-127, and Lowery, *The Spanish Settlements,* II, 258, 286.

[198]Richard Hakluyt wrote Ralegh from Paris on December 30, 1586, that the map ordered to be made by André Homem, "the prince of the Cosmographers of this age," would soon be delivered. Homem was cosmographer to Charles IX, and his nomenclature may have been French. Quinn suggests that this map may have been taken on White's expedition in May, 1587; if Ralegh received it, it would have been accessible to both White and le Moyne. Quinn, *The Roanoke Voyages,* I, 493.

[199]The large projecting peninsula between R. de May (St. John's) and R. des Dauphins (St. Augustine or Matanzas) is unlike any feature shown on any other known map, colonial or modern, though it bears a slight resemblance to le Moyne-de Bry 1590. The strong projection at C. Canaveral is a feature common to many maps of the period.

[200]Minor differences between Fernándes's and White's maps may have resulted for several reasons: White's or Dee's copyist's variations in copying the original; Fernándes's own additional contributions to his map from his knowledge of the Caribbean; or White's inclusion of additional names from other available maps.

80

[201] The latitudes and distances on the White general map, followed by the actual latitudes and distances in parentheses, are: Cape of Florida 25° (25°) to Port Royal 34°45' (32° 15') = 672 (500) miles; Cape Lookout 34°30' (33°51') to Cape Henry 40°45' (36°56') = 430 (214) miles.

[202] Quinn, *The Roanoke Voyages,* I, 461.

[203] The official Spanish charts of the Casa, on the other hand, greatly exaggerated the length and minimized the latitude change of the New England-Nova Scotia coasts.

Examples of maps containing errors similar to that of the North Carolina-South Carolina coast on White's map are: Vespucci, World, 1526, New York, Hispanic Society of America; Antonio Millo, North Atlantic Coasts, before 1590, Rome, Bibl. Vittorio Emanuele; Harmen Janz, North Atlantic Coasts, Portolan 32, 1604, Firenze, Bibl. Naz.; Joannes de Laët, "Florida et Regiones Vicinae," *Novus Orbis,* 1630; Nicholas Comberford, Chart of North America, 1650, G213.2/2, Greenwich National Maritime Museum, Greenwich, England; Hendrick Don ker, "Pascaerte vande Caribiscae Eylanden," K P/N J. 11445 M, ca. 1665, Amsterdam, Univ. Bibl.

[204] British Museum, P. & D., 1906-5-9-1(3). Quinn, *The Roanoke Voyages,* I, 461, II, 847-872, analyzes the various features of the map and gives an annotated list of all the names; Hulton and Quinn, *The American Drawings of John White,* I, 56-57, 136, has the best technical description and commentary on its construction; Cumming, *The Southeast in Early Maps,* 16-17, 120-121, includes a bibliography and examines the map's influence on later conceptions of the coast.

[205] This does not exclude the possibility that the map incorporates information gained from the first voyage of 1584 or even during White's month-long stay in 1587. The date [1586?] rather than [1585] is given, inasmuch as "Chesepiuc" and "Skicoac" are tribes unknown before the first colony. The sparseness of scenery and of known geographic information may be more explicable if the map was prepared for a politically active recipient such as Walsingham.

[206] Quinn, *The Roanoke Voyages,* II, 847-848.

[207] Reproductions in Cumming, Skelton, and Quinn, *The Discovery of North America,* 259 (Smith), and 271, 287 (Champlain).

[208] Hulton and Quinn, *The American Drawings of John White,* 54: "If they corresponded to those instructions, they were drawn from a survey by astronomical observations, employing a uniform scale and fixed representational conventions, and fitted together by a carefully defined system of registration and adjustment." For Bavin's instructions, see Quinn, *New American World,* III, 243-245.

[209] Both were versatile: Harriot drew an important map of Guiana for Ralegh, and White's coastal profiles and drawings of fortifications in Puerto Rico demonstrate an architect's eye, or at least rapid absorption of other skills. Obviously, on many subjects such as flora, fauna, and native activities and society, their disparate skills as scientist and artist reinforced each other.

[210] Quinn, *England and the Discovery of America,* 255-257, 259, 261-263. White might possibly have made a short expedition by pinnace through Currituck Inlet, which is shown on both of the White maps and by de Bry, or through Trinity harbor.

[211] Title: Americae pars, Nunc Virginia dicta, primum ab Anglia inuenta, sumtibus Dn. Walteri Raleigh, Equestris ordinis Viri Anno Dñi M D LXXXV regno Vero Sereniss: nostrae Reginae Eisabethae XXVII Hujus vero Historia peculiari Libro discripta est, additis etiam Indigenarum Iconibus [cartouche, top right corner]: Auctore Joanne With Sculptore Theodoro De Bry, Qui et excud [cartouche, left center], in Thomas Harriot, *A briefe and true report of the new found land of Virginia* (Francoforti ad Moenum: T. de Bry, 1590). Plate 1, Volume 1, of Theodorus de Bry, *Collectiones Peregrinationum in Indiam Occidentalem* (Frankfort: T. de Bry, 25 parts, 1590-1634). Quinn, *The Roanoke Voyages,* I, 462, 849-850, 852-872, is valuable for background, description, and toponymy; Hulton and Quinn, *The American Drawings of John White,* I, 54-55, 139, provides analysis of cartographic methods involved; Cumming, *The Southeast in Early Maps,* 16-17, 120, 122-123, gives the general background and bibliographical references. A first state of the de Bry map spells the Indian settlement immediately south of Chesapeake Bay "Ehesepiooc," and this spelling is followed by some early map makers, such as Corneille Wytfliet ("Florida" 1597) and Jean Matal ("Florida" 1600).

81

The second state shows a crude attempt with a burin to change the *E* to a *C*; this state, with the *C* not concealing the *E*, is shown in plate 25.

For full cartobibliographical information on southeastern and local maps (size, scale, volume location, insets, and so on), see individual map descriptions in Cumming, *The Southeast in Early Maps.*

[212]*Quinn, The Roanoke Voyages,* I, 414 n. 5, 430, n. 4. The notes were translated back into English by Hakluyt for the English edition. De Bry published this first volume of his *Grande Voyages* or *Travels into the West Indies* (hence usually called *America*) in four languages—Latin, French, German, and English—within ten days. There formerly arose some question concerning the identity of the artist because all editions except the English refer to him as "With" on the title page, as well as in all states and copies of the engraved map. (De Bry metathesized *h*'s and *th*'s frequently.) However, in the English edition translated by Hakluyt is de Bry's statement that he created his "trve pictvres out of the verye original of Maister Ihon White an English paynter," and he refers to White's "aborted" (and then-current) trip of 1588. White's authorship is no longer seriously questioned: Cumming, *The Southeast in Early Maps,* 70, n. 42. The map is found in three slightly altered states: in the first state, "Chesepiooc," the name given to the Indian tribe on the south bank of Chesapeake Bay, is engraved "Ehepiooc"; the commonest second state has been revised with an initial *C* cut in or over the *E*. In the rarer third state, *C* replaces the *E* entirely. The map remains unchanged otherwise in all editions of the book.

[213]Quinn, *The Roanoke Voyages,* II, 850; Hulton and Quinn, *The American Drawings of John White,* I, 55.

[214]Harriot, *A briefe and true report of the new found land of Virginia* (Frankfort: T. de Bry, 1590), plate II. The title of the map is printed below the plate with the accompanying explanatory legend by Harriot. It varies with the different language editions, unlike the title of the preceding map, which is engraved on a cartouche on the plate. Quinn, *The Roanoke Voyages,* I, 413, II, 853-864, reprints legend and analyzes toponymy; Cumming, *The Southeast in Early Maps,* 123-124, has description and bibliography, with the titles in all four languages.

[215]Theodor de Bry, *America,* Part II (Frankfort, 1591), plate V, "Galli ad Portum Regalem pervenum." Hulton, *The Work of Jacques Le Moyne,* II, plate 97; Stefan Lorant, *The New World* (New York: Duell, Sloan, and Pearce, 1946), 45.

[216]For a full discussion of the problems involved, see Quinn, *The Roanoke Voyages,* I, 413-415.

[217]Helen Wallis, "The first English globe: a recent discovery," *Geographical Journal,* CXVII (1951), 275-290. Dr. Wallis found the globe at Petworth; it derives from the estate of the ninth earl of Northumberland, who was closely associated with both Ralegh and Harriot. See also Quinn, *The Roanoke Voyages,* II, 850-851.

[218]The former ascription of this map, *Novvs Orbis,* to Francis Gaulle is considered unfounded; the identification of the initialed signature as that of the Flemish engraver Filips Galle (F[ilips] G[alle] S[alutem]), suggested by Quinn and Skelton, is now generally accepted. (Hakluyt, *Principal Navigations,* xlviii, n. 1.) For comments and reproduction, see Cumming, *The Southeast in Early Maps,* 121, plate 9; and *The World Encompassed* (Baltimore: Walters Art Gallery, 1952), No. 203, plate XLVI.

[219]Cumming, *The Southeast in Early Maps,* 121-122.

[220]Cumming, *The Southeast in Early Maps,* 122.

[221]Cumming, *The Southeast in Early Maps,* 96, plate 16.

[222]Cumming, *The Southeast in Early Maps,* 127-128, plate 17 (Florida and Apalche); A. E. Nordenskiöld, *Facsimile-Atlas to the Early History of Cartography, with Reproductions of the Most Important Maps Printed in XV and XVI Centuries.* Translated by Johan Adolf Ekelöf and Clements R. Markham (Stockholm: P. A. Norstedt and Söner, 1889; reprinted, with corrections and new introduction by J. B. Post, New York: Dover Publications, 1973), plate 1 h.

[223]Cumming, *The Southeast in Early Maps,* 128.

[224]"Virginia Item et Floridae Americae Provinciarum, nova Descriptio," Gerard Mercator, *Atlas . . . auctus ac illustratus á Jodoco Hondio* (Amsterdam, 1606), No. 143. Cumming, *The Southeast*

in Early Maps, 17-18, 129-130, examines the construction and influence of the map and gives references and bibliographical information. A list of regional maps that follow de Bry and Hondius is given in Cumming, "Geographical Misconceptions of the Southeast," 487-489. By the middle of the seventeenth century the White topography and toponymy had generally spread to continental and world maps.

[225]The original Cape Fear is here misspelled "C. of faire id est, Prom. tremendum," an error copied on many later maps that may have contributed to the shift of the original name to a different cape, the present Cape Fear.

[226]Cumming, "Geographical Misconceptions of the Southeast," 478-486.

[227]Simancas, Spain. Archivo General de Simancas, M.P.D. IV-66, XIX-163. Cumming, *The Southeast in Early Maps,* 131-132, plate 21 (detail); Alexander Brown (ed.), *The Genesis of the United States* (Boston: Houghton, Mifflin, 2 volumes, 1890), II, 184-185, with map; Philip L. Barbour (ed.), *The Jamestown Voyages under the First Charter, 1606-1609* (Cambridge, England: Hakluyt Society, second series, 2 volumes, 1969), I, 138-140, with map (best reproduction). The chart is anonymous; Captain John Smith is the most probable source. See also David Beers Quinn, *Set Fair for Roanoke: Voyages and Colonies, 1584-1606* (Chapel Hill: University of North Carolina Press for America's Four Hundredth Anniversary Committee, 1985), 371-372 (reproduction) and 439, n. 25.

[228]Quinn, *The Roanoke Voyages,* I, 264.

83

Appendix

**Spanish Cartographic Nomenclature of the
Southeastern North American Coast in the Sixteenth Century**

The following table enumerates the sixteenth-century names for the east coast from the Cape of Florida to Cape Cod. The names derive chiefly from the voyages of the Licenciado Lucas Vazquez de Ayllón or his captain, Pedro de Quexós (1521-1526), and of Estéban Gómez (1524-1525) as recorded by cartographers and cosmographers of the Casa de Contratación in Seville and as these names were added to or modified on significant maps later in the century. The table follows the variations in nomenclature, spelling, punctuation, etc., of each map; it does not attempt to solve the complexities of toponymic identification along the southeastern coast of North America. Except for the two Weimar maps, the toponymy was transcribed by the writer from the original charts; the list has been checked against photographs of the originals. An attempt has been made to ensure accuracy.

A list of modern names and landmarks with exact latitudes is given for convenience. No attempt is made here to correlate every Spanish name with its modern equivalent. The list extends both north and south of the North Carolina coast and beyond the territorial limits of this work. Nomenclature of the southeastern coast is analyzed, with possible identifications, by Woodbury Lowery in *The Spanish Settlements within the Present Limits of the United States* (New York and London: G. P. Putnam's Sons, 2 volumes, 1901, 1905); in John R. Swanton, *Early History of the Creek Indians and Their Neighbors,* Bulletin 37, United States Bureau of American Ethnology (Washington, D.C.: United States Government Printing Office, 1922); and in various editions of original texts, though many of their conclusions have been disputed or corrected. Pertinent names relating to the present work are examined in the text for probable identification. Variations even in the maps closest to the padrón general in Seville exhibit disconcerting differences arising from the attempt by the official cartographers and cosmographers in Seville to correlate the information received from pilots' sketches and navigators' logs and reports. It is frequently impossible to provide accurate or certain identification of the landmarks given. Names are entered in different order and in different locations, and omissions, confusing misspellings, and great variation in the delineation of the coast are common. Unlike the West Indies, where trade, population, and courts of law provided uninterrupted documentation, and Canada, where Portuguese and French fishing and trading activity was frequent and attempts at colonization were reported, the southeast coast for more than fifty years provided little opportunity for cartographic confirmation, analysis, or change of the original information brought back to Europe from three brief voyages in the 1520s.

Spanish Coastal Names from Florida to Cape Cod on Sixteenth-Century Maps

Modern Toponymy At Given Latitudes	Vespucci 1526	Ribero-Weimar c. 1527	Ribero-Weimar 1529	Ribero-Vatican 1529	Wolfenbüttel c. 1530
Cape of Florida	C: anito			La Florida	namico [?]
	[Cape of Florida]		R. salado		R. Salado
Cape Canaveral 28°28′			c: roxo		C. Roxo
Mosquitos 29°2′	Trā floryda				R. de corrientes
St. Augustine Inlet 29°53′30″					R. Florido
St. Johns R. 30°24′		aguada			R. solo
St. Mary's R. 30°42′			marbaxa	marbaxa	arreciffes
St. Simon Sd. 31°7′30″	Trā nueua			R: solo	marbaxa
Altamaha R. 31°19′	de ayllon		c: gruesso	C. gruesso	R. solo
Savannah R. 32°02′					C. gruesso
Hilton Head 32°13′	C. da sata elena				
Port Royal 32°15′	R. de La + (40°N)		C. de S: elena	C: de S. elena	C. de S. elena
St. Helena Sd./S. Edisto R. 32°27′		tierro del	S: elena	S elena	S. elena
N. Edisto R. 32°33′30″		licençiado ayllon		playa	
Charleston Bay 32°45′	R. Jordan		R: Jordan	R: Jordan	R. Jordam
Cape Romain 33°	C. da sa njcolas				
Santee R. 33°08′	r de arezifes				
Winyah Bay 33°12′30″			C. de S: Romā	C: de s romā	C. de S. Roman
North Island Point 33°12′30″				R: de canās	
Cape Fear 33°51′	R da sa terazanas			R: del sucio	R. sucio
New Inlet 34°31′30″			R. del principe	R: del principe	R. del principe
Bogue Inlet 34°38′					Arholadas [?]
Cape Lookout 33°35′	C de Trafalgar		C: traffalgar	C: traffalgar	C. tr a Falgar
Ocracoke Inlet 35°04′					
Cape Hatteras 35°14′					
Oregon Inlet 35°46′50″					
			R. d'l espū stõ	R: del espū s	R. del spu sto
Chesapeake Bay C. Henry 36°56′	baya de sa ma		b. de S. m^a	b. de S m^a	b. de s: Maria
James River 37°	[40½°N.; Azores, 38-41°N]		[34°-35°N]		Tiera del licenciado aillon
Cape Charles 37°6′	C. de sa ma				
			playa	playa	playa
Cape Henlopen 38°48′			C: de S: Juan	C: de Juā	C. de S. Juhan
	R. de s Juhan				
Monomoy Point 41°32′30″			R. destīago		R. de S. tiago
Cape Cod 42°04′57″		C de arenas	C: de arenas	R: de stīago	C. de arenas

Alonso de Chaves Qvatri-partitv ... Espeio de Navegantes [1539] fols. 41v to 44	Lopo Homem 1554	Diego Gutierrez 1562 [ante 1554]	Alonso de Santa Cruz c. 1567	Fernandes-Dee 1580	Dee 1580
THE EAST COAST OF FLORIDA	Tera Florida	La florida	florida	La florida	Terra Florida
East point of Martires 25°50 leagues N¼ NE to		costa de fuego	tierra del quasto	costa de fuelgo	The Bahama Channel
c. canaueral 27½° 20 l. NW to	C. del canaueral	C de Cannaue	c. del canaueral	C. del camaueral	flows ever to the North
		Rio de corintes	r. de moschites		C. de Canaveral
Rio de Corientes 28½° 15 l. NNW to	C. de cruz		r: de s. agostino	C de Corientes	
Cabo de cruz 29⅓° 24 l. N to		C. de aruz	r: de s: matte	R. de Erne	C de +
marbaxa 30¾′ 20 l. NNE to C. grueso		marbaxa	r: de s: ioan	Marbaxo	Marbaxa
This is a kind of bay with some shoals		R: de terra llana	r: de s: age		
Rio seco 31¼° [10 l. to]			r: de s: andres		R. secco
Cabo grueso 31⅔° 20 l. NE to	C. Grueso	c. gruedo	r: de guales	Tiera lhana	
	arenas		r: de tatatecoro		Costo brava
Cabo de santa elena 32⅓° 30 l. NE to C. Roman	C. de Santa Ileña	C. de S. Elena	r: de balenas	C. de S. Elena	R. de S. Ellena
West of this cape is a large bay	R. de Sta Ileña	R. de S. Elana	r: de s: elena	R. de S. Elena	
Rio de santa elena 32⅔° 30 l. NE to C. Roman	ancones		C: de s: elena	C. Amcon	
Rio Jordan 33½° 4 l. E to	R. Jordan	Andenes	r: de medanos	Rio Jordan	R. Jordan
This large river is called Rio de aillon	c. de San Roman	R. Lordan	r: dulce		
because here he died		c de S. Roman	r: de guamas	C. de S. Tomas	
Cabo de sant Roman 33½° 22 l. NE to	R. de canoas		r: de balenas	R. de cunaas	R. de Canoas
In front, shoals extend out to sea about 3 l.	R. de boca	R. de canoas	r: de canoas	ancones	R. Baxo
On the west shore is a cove and Jordan R.	R. frio	Ananes	r: de nogales	R. seco	P: del Principe
Rio de canoas 34½° 60 l. E ¼NE to Trafalgar	R. de las baxas	Terra lhana	r: de pinos	Tiera lhana	
Rio de los baxos 35° 22 l. E. ¼NE to	R. del princepe	R. baxo	r: deloro		
Rio del principe 35½° 14 l. E to	c. de traffalgar	P: del principi	c: de s: romano	C de Traffalgar	C de Terrafalgar
Cabo de trafalgar 35½° 20 l. N to		c. de terrafalgar	r: del principe		
This cape extends outward most on this			c: de trafalcar		
whole coast and is the best known.					R. Spirito S.
THE NORTH COAST OF FLORIDA					
Baya de santa maria 36½° 20 l. NE to C. de sant Juan				B. de S. Maria	G. de S Mª.
A large bay with many islands in it; it is	C. del spirito sto	R. de santo spirito	r: del spto: santo		
the entrance to two rivers	C. de Salado	Bª DE S. MARIA	bª: de nra senora	R. Saluador	
Rio de espiritu santo 36½° Entrance SW to bay		R. salado	r: salado		R. Salada
Rio salado 37° Entrance to N in the bay	praia	R de la playa	bª de enganno		
Cabo de sant Juan 37° 24 l. NE ¼ N to	C de S joan	C. de S. Ivan	c: de s: joan	Plaia	C. de S. John
	C de las	C. de las Arenas		c: de S. Jo.	C. de S. tiago
Cabo de las arenas 38⅓° 30 l. N to	Stiago			C de las Areans	C. de Arenas
Cabo de Santiago 40°					
Most prominent cape on this entire coast			bª de s: Xpobal		
To its W the bay de Xpoual, a great bay	b. de san cristoval				

PLATE 1.
Juan de la Cosa. World Chart. 1500. Madrid, Museo Naval.

This map, signed and dated by Cosa, who sailed with Columbus in 1493-1494 and made three later voyages, is drawn with pen and ink and watercolors on oxhide. To the north, with five English standards, is the legend "the sea discovered by the English," which refers to Cabot's voyage of 1497, during which he reached North America, possibly along the coasts of Newfoundland and Nova Scotia. To the south is the latest information recorded by the Spanish in the West Indies and South America. Between, lying in a southwesterly direction, is a coastline that may represent an attempt to show Asia, although no names or legends support this theory. It may demonstrate knowledge of a voyage along the North American coast; conclusive evidence, however, of such explorations by surviving ships of John Cabot's 1498 expedition or by other vessels is lacking.

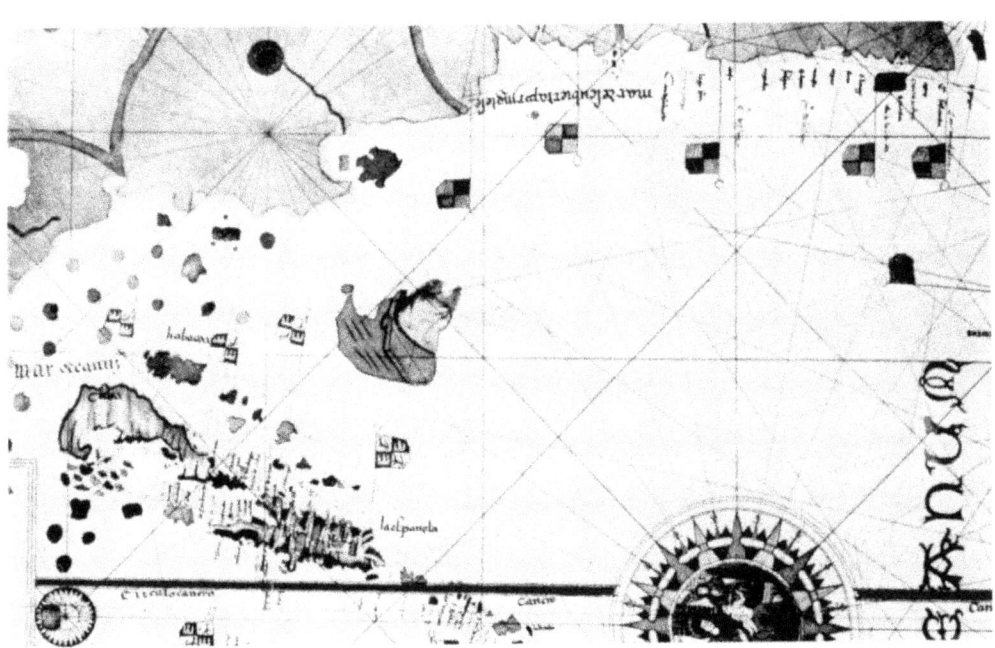

This planisphere was brought from Portugal to Ercole d'Este, duke of Ferrara, at his behest by his envoy, Alberto Cantino. The most baffling problem on the Cantino map and on a whole series of later related maps is the large unnamed landmass to the northwest of "Isabella," presumably the island of Cuba, that stretches northward to beyond 50° N.L. Some scholars claim that the landmass is a reduplication of Cuba; others vigorously defend the theory that it is based on a chart of an unrecorded discovery of Yucatan, placed in-correctly and at right angles to its true position.

A third suggestion is that it delineates Florida and lands to its north, with the Key West islands to the south and the east coast of the peninsula cor-rectly trending north by north-west. On the basis of present evidence, this suggestion seems to be the most reasonable. To the objection that there exists no record of a voyage along the southeastern Atlantic coast before 1502, it may be said that unlicensed voyages were being made and that records indicate that full reports of some known voyages are lacking.

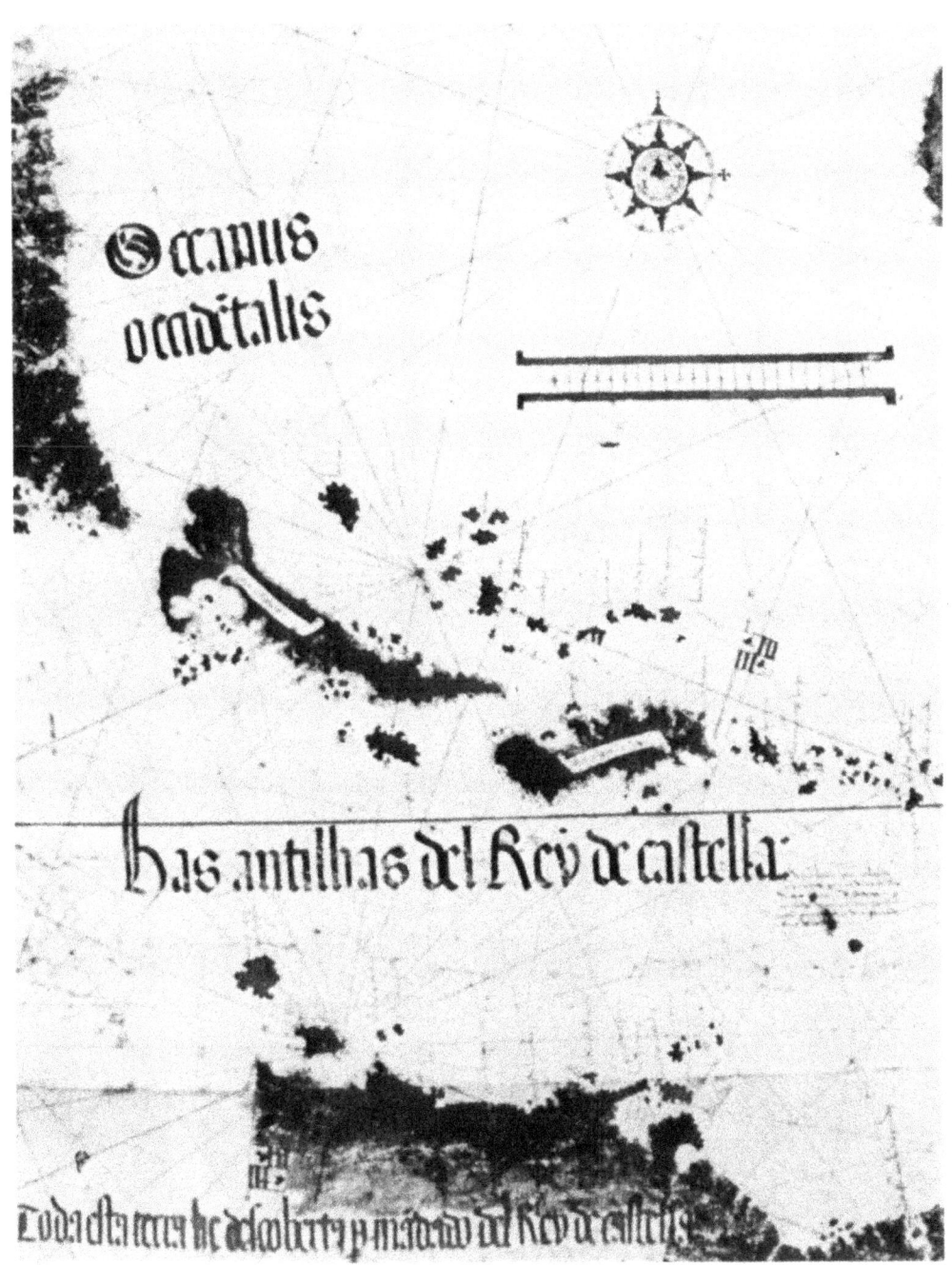

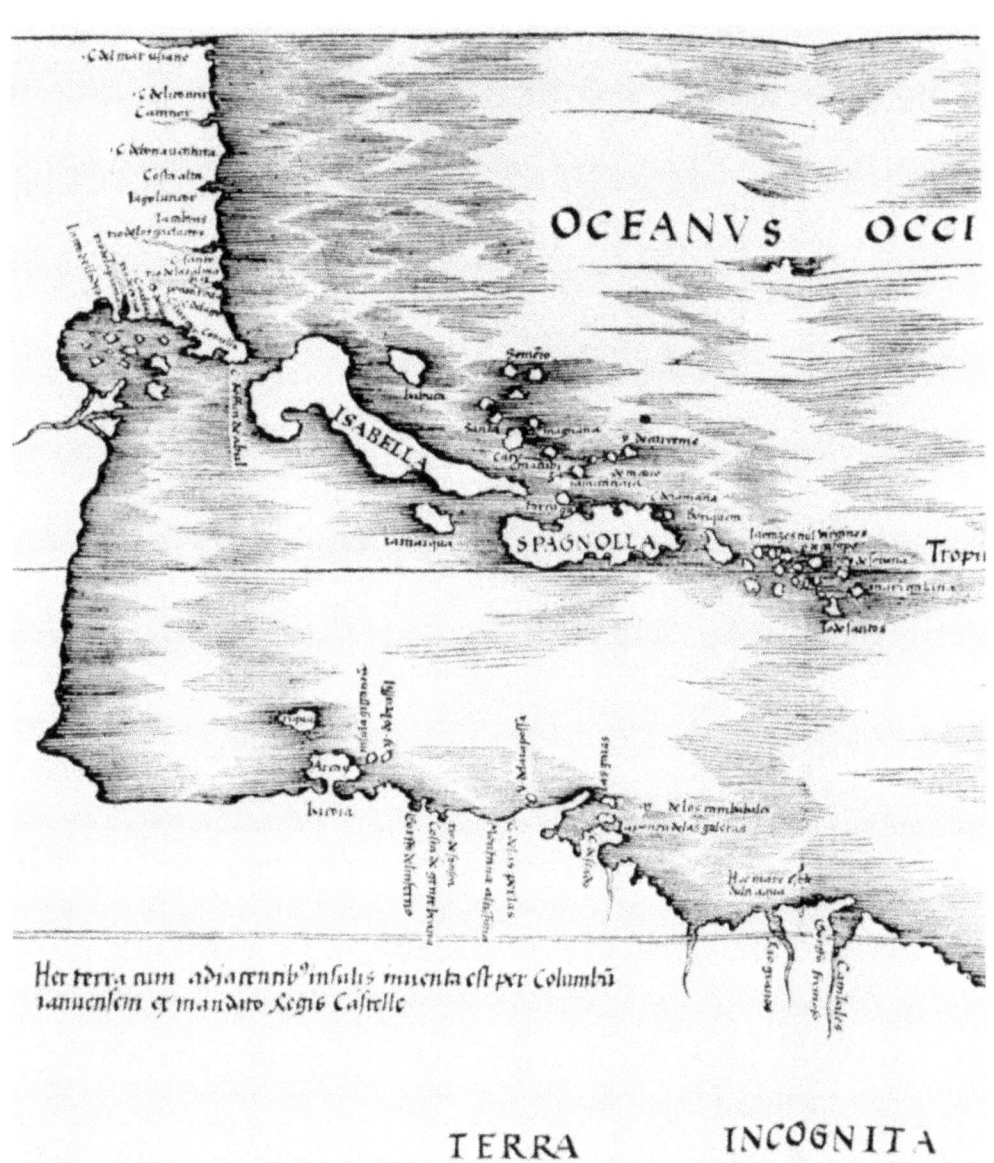

93

PLATE 4.

Juan Vespucci. World map. 1526. New York. Hispanic Society of America.

This detail from a map-pamundi made in Seville by Juan Vespucci, nephew of Amerigo Vespucci, records the voyage of discovery along the southeastern coast in an expedition sent by the Licenciate Lucas Vasquez de Ayllón under Pedro de Quexós in 1525. Along the coast are names given by Quexós to the "New Land of Ayllón," recently granted by Charles V of Spain in a royal çedula to Ayllón, a wealthy auditor of the island of Santo Domingo. A previous voyage sent out by Ayllón had discovered the unknown shores in 1520.

The names here recorded became established nomenclature for the region on most Spanish maps through the century and beyond, although more careful assessment of reports soon changed and improved the coastline and added to the toponymy. The name "Elena" is still in use in St. Helena's Sound, South Carolina. At the northern limit of Ayllón's territory is Cabo de Santa Maria and the Baya de Santa Maria, the North Carolina sounds and probably also Chesapeake Bay. Above Santa Maria the coast ends, and a great ocean gap is shown between it and Terra de los Bacallaos. The reports of the Gómez voyages of 1524-1525 along the New England coast had not yet reached Seville.

Juan was the son of Antonio, Amerigo Vespucci's oldest brother; after the death of Amerigo, the pilot major, in 1512, Juan Vespucci was appointed pilot of the Casa de Contratación and in 1515 became a pilot to the king and a reviser and maker of copies of the padrón real, the great official map kept in the Hydrographic Department of the Casa in Seville, on which were recorded new discoveries reported by pilots returning to Spain.

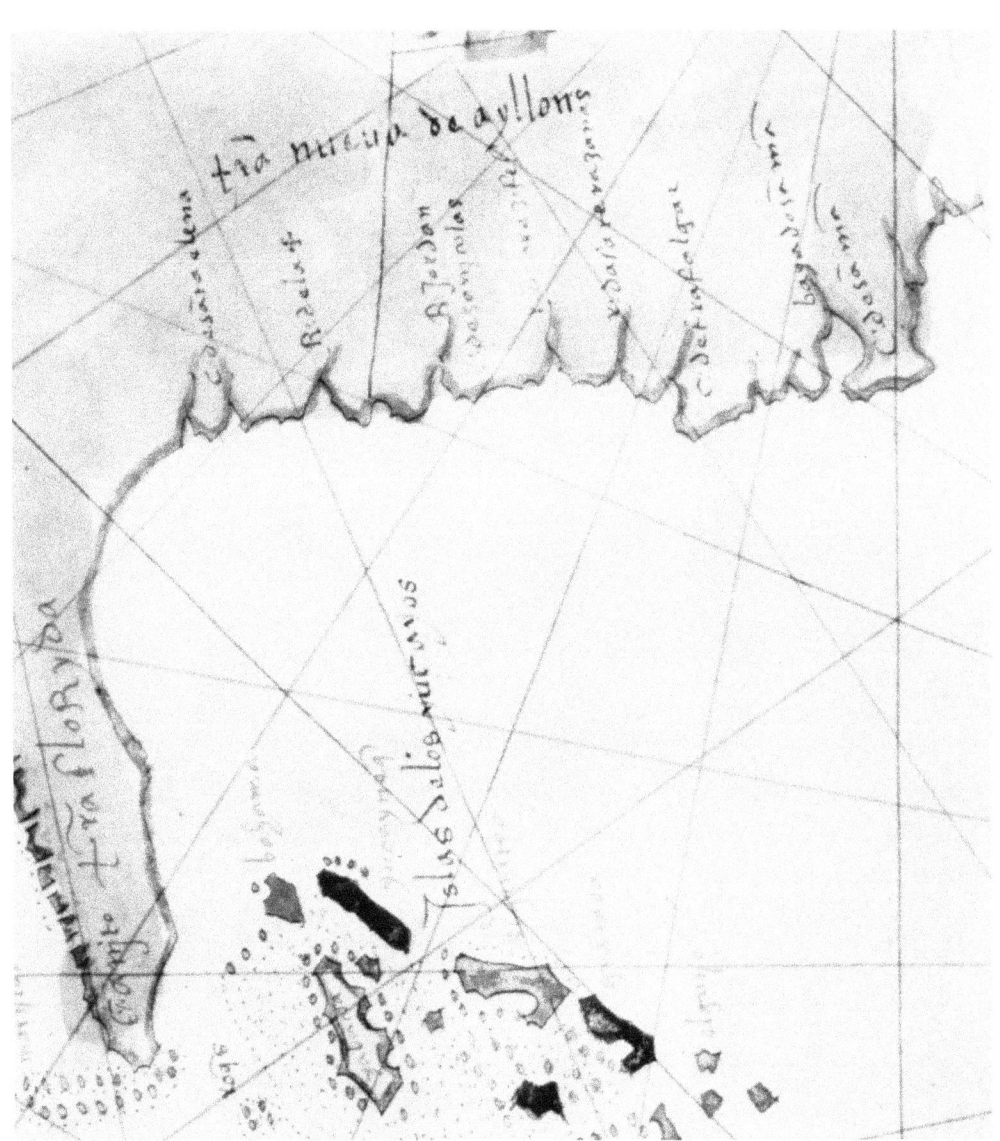

95

PLATE 5.

**Gerolamo da Verrazzano.
World Map. 1529. Rome.
Biblioteca Apostolica
Vaticana, Borgiano I.**

Giovanni da Verrazzano, sailing along the Carolina coast in 1524, thought that the narrow strip of Outer Banks islands was a 200-mile-long isthmus separating the Atlantic and Pacific oceans. Verrazzano was a Florentine in the service of Francis I of France; upon his return he wrote a letter to the king describing his voyage from Carolina to Nova Scotia and referring to the isthmus as being one mile wide. The map drawn in 1529 by Gerolamo, Giovanni's brother, includes many names not mentioned in the letter and gives the width of the isthmus as six miles.

Throughout the century the geographical error of the isthmus and the indicated closeness of the Pacific excited Europeans searching for a westward passage to the Spice Islands and Cathay; a map and a globe presented by Verrazzano to Henry VIII of England and showing Verrazzano's Sea or Bay was one of the stimulating factors in the Englishmen's choice of a location for the Ralegh colony. Many Verrazzanian-type maps were made, but the Spanish maps were dominant, even in France.

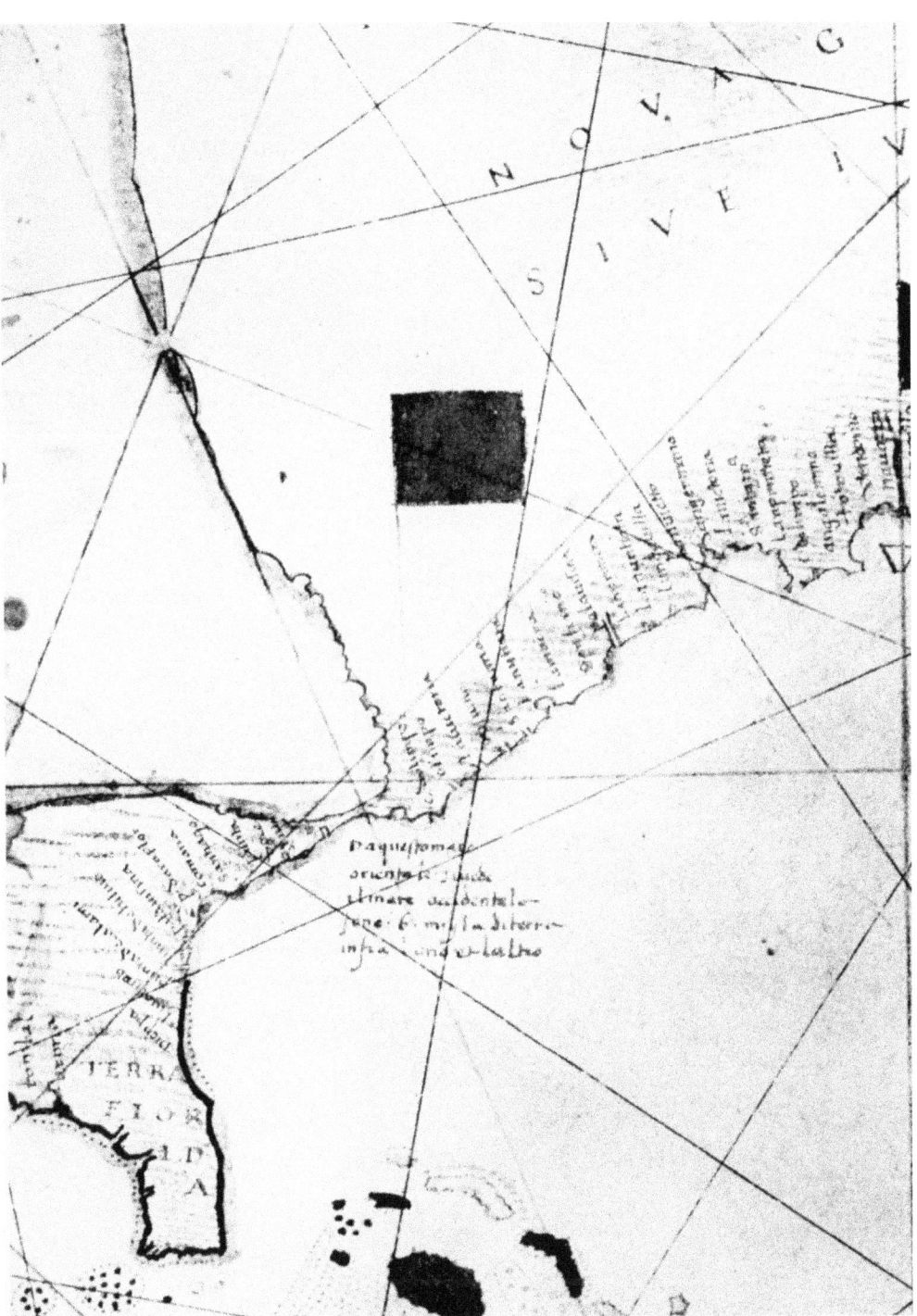

97

Ribero (also spelled Diogo Ribeiro), a Portuguese pilot and an expert maker of charts and nautical instruments, was attracted to Seville and became a valued official in the hydrographic department of the Casa de Contratación, the powerful board of trade with extensive powers over foreign policy.

As early as 1525 unsigned charts often attributed to Ribero were ordered by the king as presents to allied sovereigns. These handsomely designed maps, showing the latest Spanish discoveries, included the current information provided the padrón general (called padrón real before 1526), the official chart kept at the Casa.

This chart, formerly in the Grand Ducal Library at Weimar, and another, presently in the Vatican Library, are both dated 1529 and signed "Diego Ribero Cosmographer to his Majesty." They both show a continuous Atlantic coastline without break by a westward passage in the Western Hemisphere; and they incorporate the reports of both the Ayllón and Gómez voyages with the geographical delineation and names that remained fundamentally unchanged on Spanish charts for more than a century.

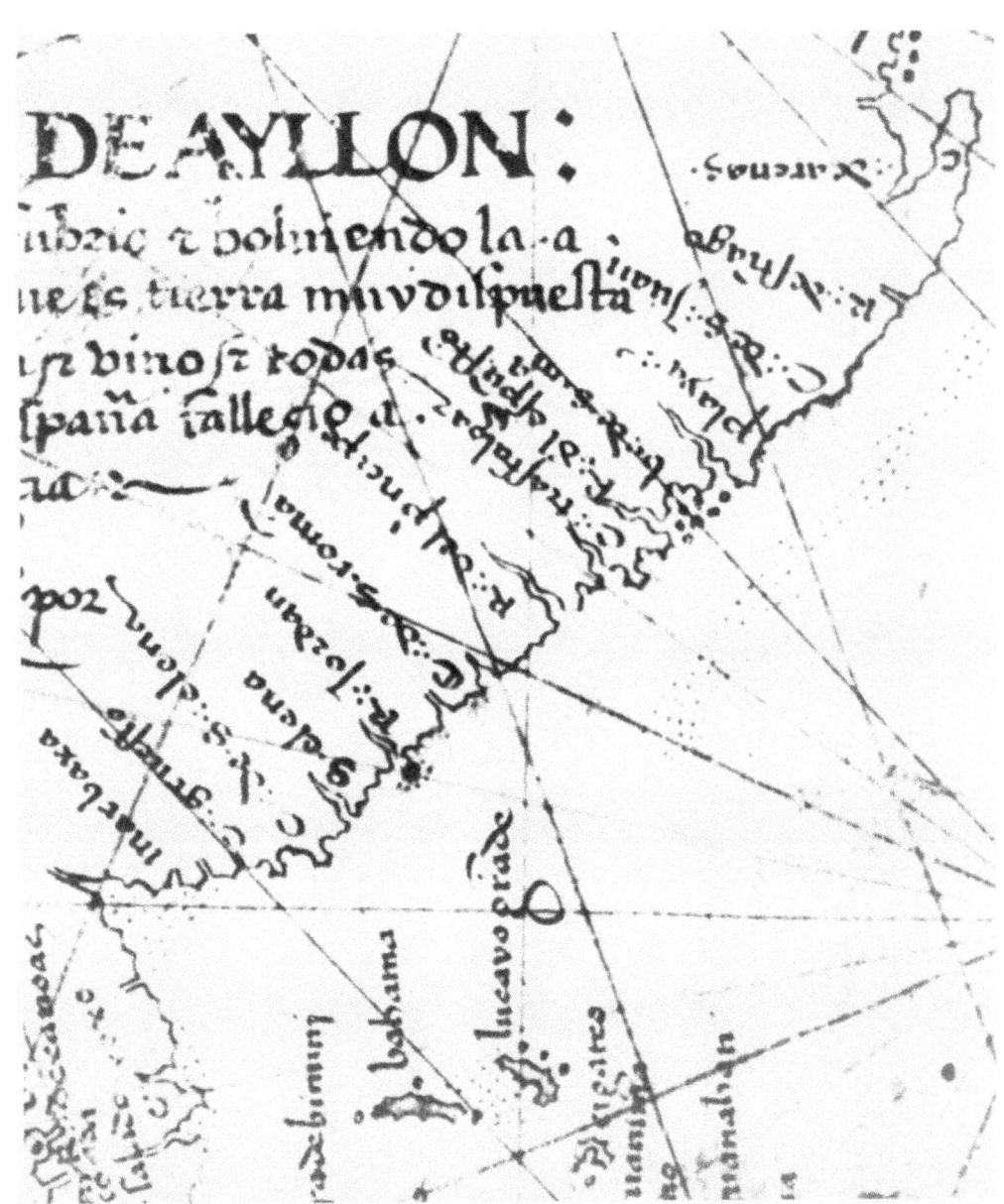

*The voyages of Ayllón and
Gómez are carefully delineated
and correlated on the two 1529
Ribero charts. The charts give
Ayllón's explorations with much
more factual information—
including that reported by survivors of the last disastrous
expedition—than was possible
for Vespucci to enter on his
earlier chart. They include
more landmarks with names;
the great open bay with
numerous islands is placed at
34°-35° (Weimar) and 36°-37°
(Vatican) instead of at 41° as
on Vespucci. This placement is
correct for the Outer Banks
and sounds and probably includes the Chesapeake. The
correlation with the discoveries
of Gómez in New England is
less accurate; any exploration
between the Chesapeake and
Cape Cod was perfunctory and
hasty, for the actual distance
between the two areas is
foreshortened by more than 400
miles, with the intermediate
shoreline almost barren of
detail.*

*This 1529 planisphere in the
Vatican by Ribero is considered
one of the most beautiful examples of cartographic art.
Larger and with more detail
than the Weimar chart, it is
more elaborately though
tastefully decorated with trees,
buildings, birds, animals, ships,
and figures. The figures, such
as an astrolabe, and legends
give a rich summary of contemporary navigational information. Along the coast from
Florida to Cape Cod, with its
pointed extremity, is a long
dotted strip, which in these
early maps is a symbol for
shoal water.*

101

PLATE 8.
Sebastian Münster. Novae Insulae. 1540. From Münster's edition of Claudius Ptolemy, *Geographia* (Basle, 1540).

This double-page map, one of the most popular printed maps of the New World in the sixteenth century, appeared in Ptolemy's Geographia *and Münster's* Cosmographia *in several languages and in numerous editions. It shows the extraordinary increase in geographical knowledge gained in less than fifty years after Columbus's discovery and some curious misconceptions of the time: Verrazzano's Sea divides North America into two continents, Francisco and Florida; South America, "the Atlantic Island which they call Brazil and America," has Amerigo Vespucci's land of cannibals with a human head and a limb being roasted, and to the south the Patagonian region of giants; Magellan's ship has passed through the strait named after him into the Pacific on its way around the world; the Pacific Ocean itself is contracted to conform to the earth's supposed 19,000-mile circumference, thus bringing close to the unexplored California coast Japan (Zipangu) and its archipelago of 7,448 islands reported by Marco Polo in the third book of his travels. The New World is shown as one continuous landmass, making the reported possibility of finding a passage to Verrazzano's Sea increasingly desirable.*

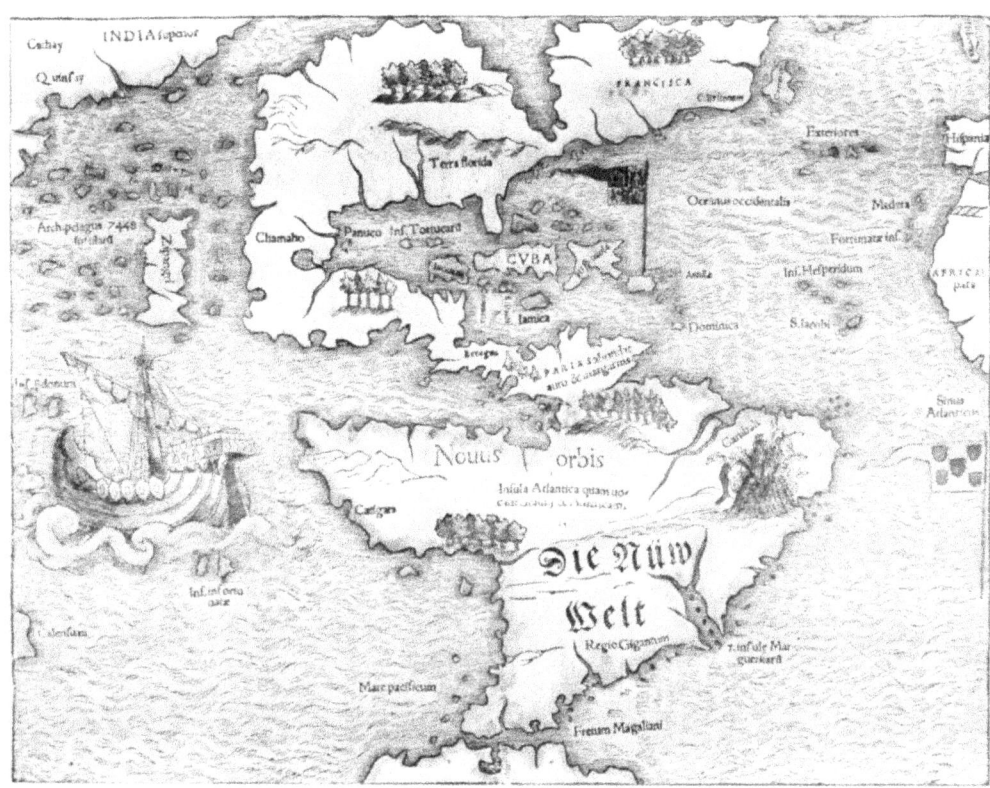

103

PLATE 9.

John Rotz. Atlantic Coast of North America. 1542. London. British Library. Royal MS 20.E.IX, fols. 23v-24r.

This map is found in a magnificently designed and illustrated atlas, The Boke of Idrography, *presented to Henry VIII in 1542 and made by John Ross, a Frenchman of Scots descent who had come to England for preferment. The work is of the Dieppe school of mapmakers, whose members enjoyed considerable maritime experience and connections. The detail here reproduced is evidently ultimately based on some padrón general, probably obtained more directly through a Portuguese chart.*

The map was in the Queen's Library, still at Westminster, and possibly was not known to the planners of the Ralegh voyages. The wide bay at 36° N.L. with its sheltering islands seems an ideal haven for preying on passing Spanish treasure fleets and for returning from raids on the West Indies.

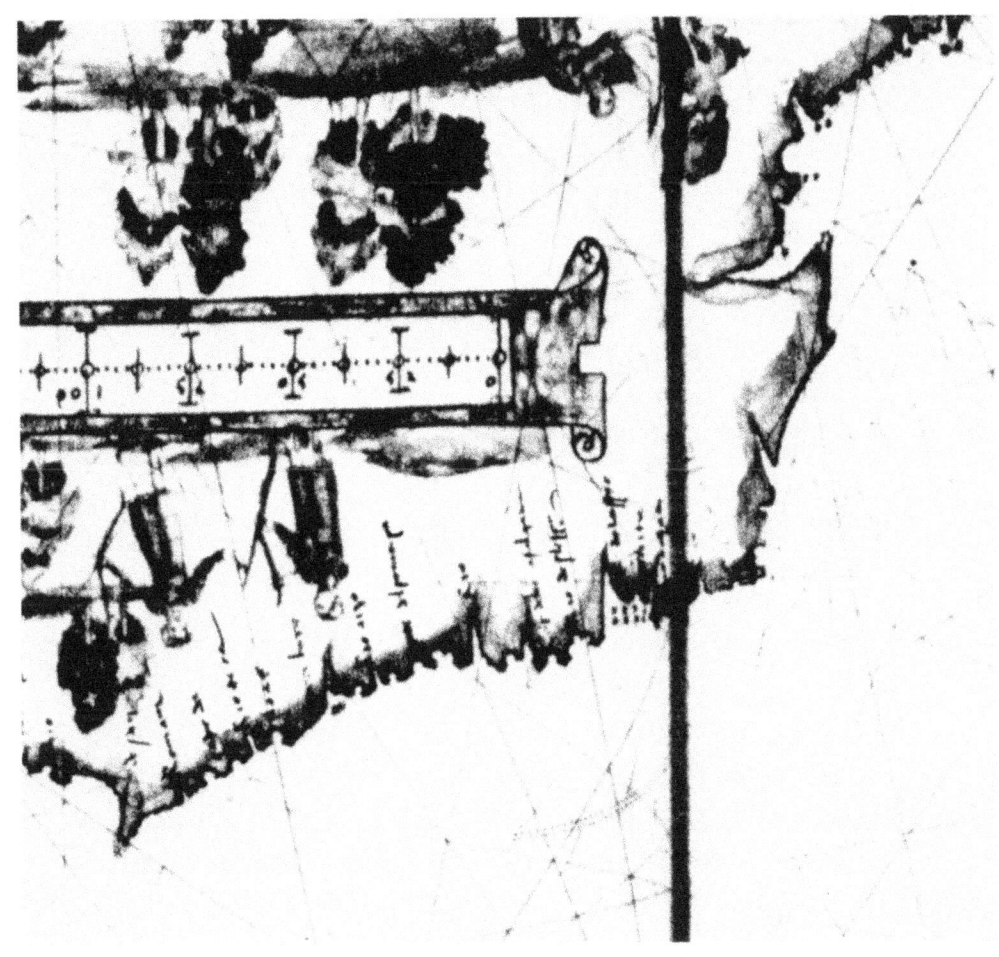

105

107

PLATE 11.

Anonymous. North Atlantic Coast of America. ca. 1545. From an atlas in The Hague. Koninklijke Bibliotheek, MS 129.A.24, fols. 25v-26r.

This atlas is of the Dieppe school, probably of Portuguese authorship with decorative and pictorial illustrations by a French artist. In nomenclature and coastal outlines this map closely follows the model established by the Spanish padrón general as revised by Alonso de Chaves in 1536 and reproduced in other contemporary maps. The northern arm of the bay "Rio Sallado" (Salt River), may be the Chesapeake, which extended north and is markedly more saline than the North Carolina sounds. The geography is almost wholly Iberian; no knowledge of Jasques Cartier's discovery of the Gulf and River of St. Lawrence in 1536 is shown; the artist's depiction of lions and a unicorn is evidently ornamental rather than an attempt to reproduce local fauna, as in most maps of the Dieppe school.

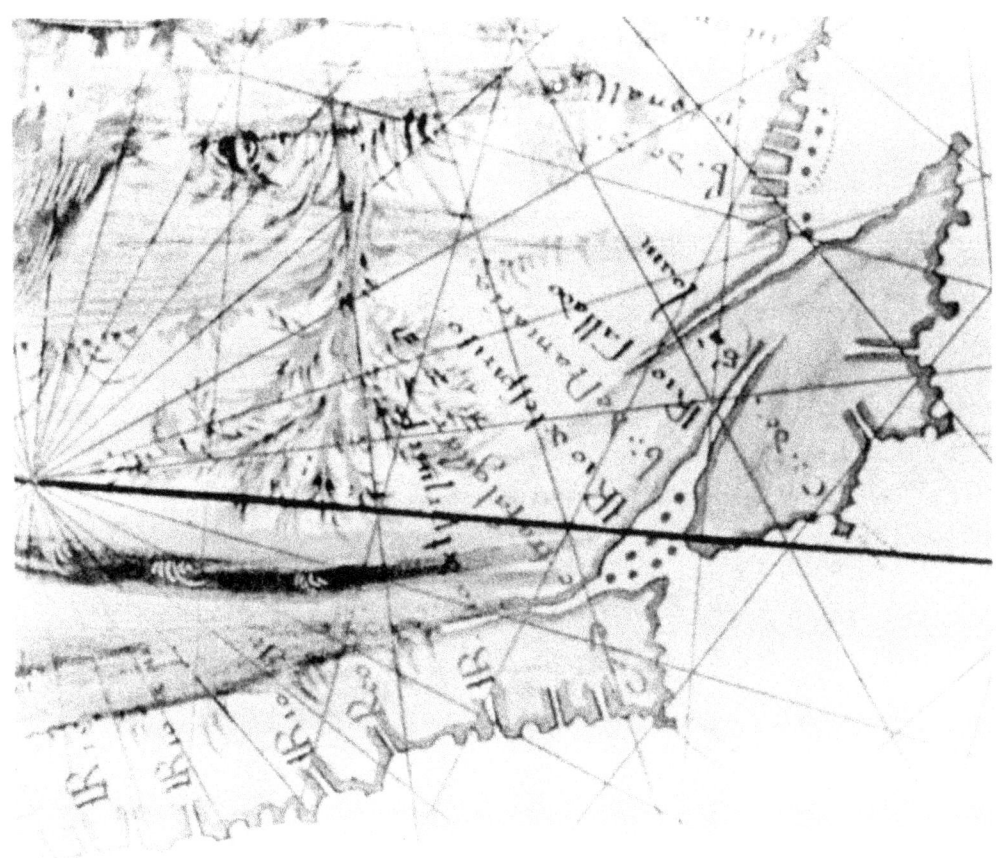

Above the Florida peninsula is a river or strait, Rio de S.^e helene (at Port Royal), that leads from the Atlantic to the "east sea" (Verrazzano's arm of the Pacific). Except for this, the coast to the Gulf of St. Lawrence is based on a Portuguese source, ultimately deriving from the Chaves 1536 padrón general. This map or one remarkably like it was used by John White in drawing his general map of the Southeast; both have a long, narrow strait at this point between two great bodies of water. There is, however, no evidence that the Ralegh colonists attempted to explore the Port Royal estuaries for a westward passageway, as Lane explored the Roanoke River.

The map on which this detail appears is an undated, unsigned, magnificently illuminated manuscript, approximately 4 feet by 8 feet, evidently of the Dieppe school, although unlike other extant examples. It derives its name from its former owners, Robert and Edward Harley, the first and second earls of Oxford, although its earlier provenance is unknown. It is also called the Dauphin Map because it includes the arms of the French heir-apparent, partly altered to royal arms.

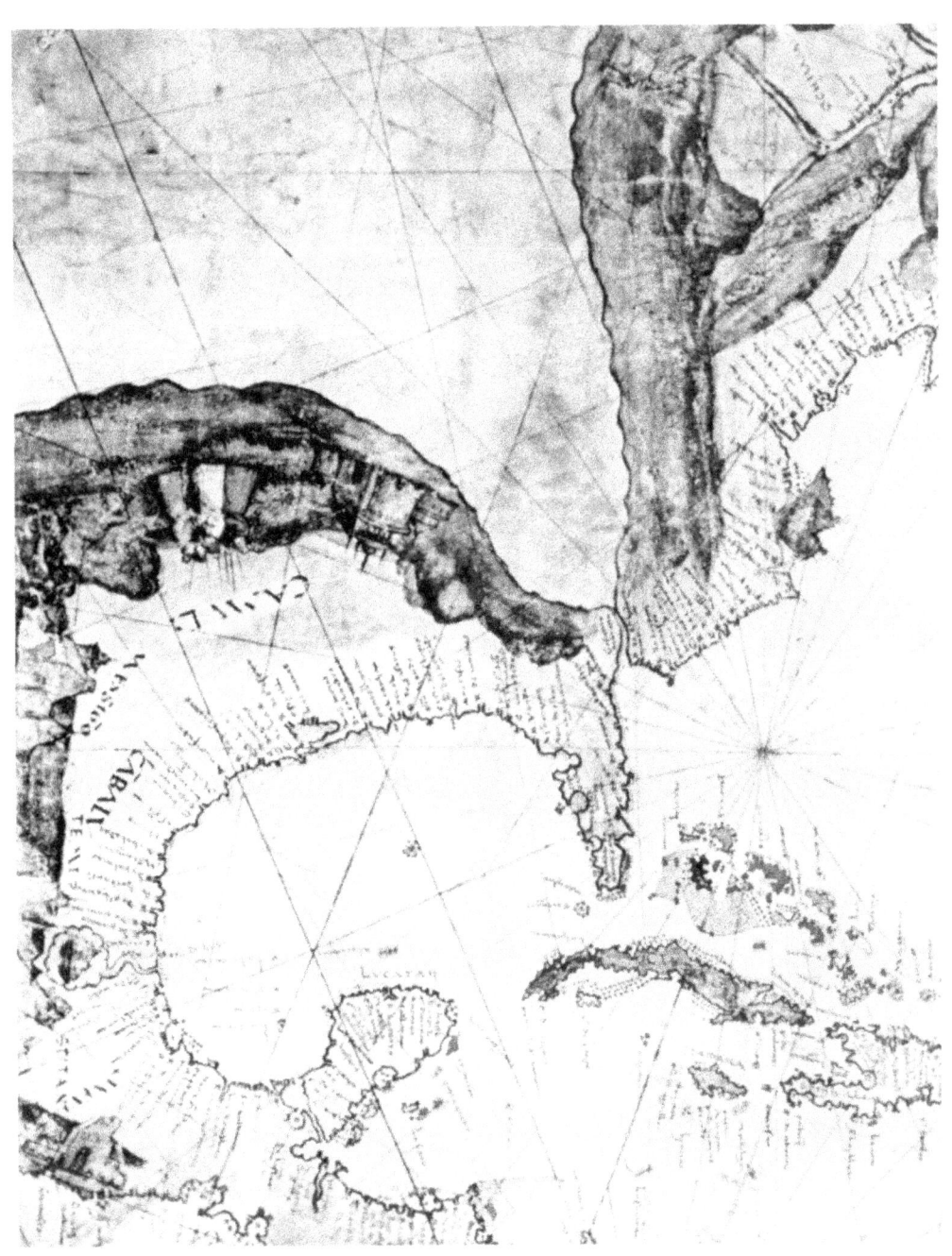

111

PLATE 13.
Pierre Desceliers. World Map. 1550. London. British Library, Add. MS. 24065.

This is a fine example of the craftsmanship of the Dieppe school, signed and dated at Arques, near Dieppe. It is based on a Portuguese chart; but unusually full nomenclature between the open, island-studded Baya de Santa Maria at 35½° N.L. and Cape Cod at 41°, with ten names instead of the usual one to three, may indicate French voyaging along this neglected coast. In the St. Lawrence River region, new details of French explorations of 1535-1542 are given, along with a long explanatory legend. This or other similar Desceliers maps may have been in England by the third quarter of the sixteenth century.

This map has at the Baya de Santa Maria a large "G. de. S.M." with a narrow entrance to the baya or gulf, which is studded with islands. The atlas in which it appears was made by a member of a celebrated Portuguese family of cartographers and was possibly available in the Queen's Library to planners of the Ralegh voyages.

It is similar to the maps of Lopo Homem, probably the father of Diogo, who had full access to all documents in the Casa da India, the Portuguese equivalent of the Spanish Casa. Diogo apparently also used the Casa in Lisbon until he fled to London after 1545 to escape from a murder charge, of which he was later exonerated. He gained such a reputation in England that he was ordered by Queen Mary to make this magnificent volume, richly adorned in gold leaf, for her husband, Philip II of Spain. Mary died in 1558, however, and the atlas remained in England.

The fact that Philip's arms have been scratched out of the map where England is shown suggests that the atlas was consulted by Elizabeth and perhaps others. It is also probable that maps by Diogo's kinsman, André Homem, whose works were similar to Diogo's, were available in England; Hakluyt commissioned maps from André Homem for Ralegh in Paris, where André was royal cosmographer.

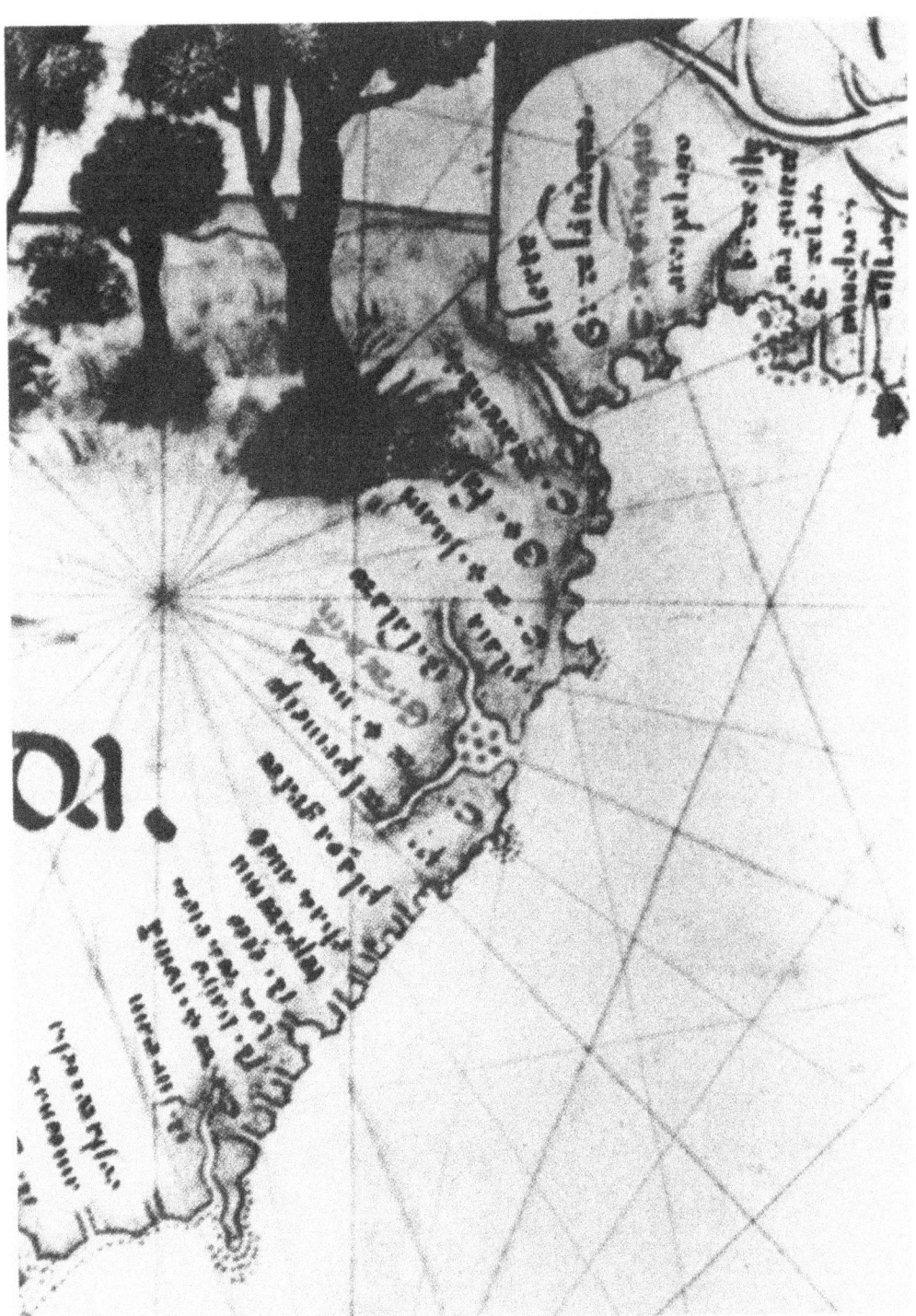

115

PLATE 15.
Diego Gutiérrez.
Americae . . . nova et exactissima descriptio. 1562.
Washington. Library of Congress.

This is a conception of North America shortly after the middle of the sixteenth century as presented by Gutiérrez, cosmographer to Charles V of Spain and acting pilot major of the Casa de Contratación during the absence of Sebastian Cabot. The section here reproduced is part of the largest and most detailed engraved map of the New World yet to be published and must have been examined with special interest and respect in England as the product of the royal Spanish cosmographer.

The map is not as authoritative, however, as it appears. Diego Gutiérrez died in 1554, and the circumstances of its publication are unknown. The Flemish engraver Hieronymus Cock was more interested in the decorative beauty of his work than in accuracy; he shows carelessness or ignorance in spelling (aruz for cruz, gruedo for grueso) and in coastal details.

117

PLATE 16.

**Alonso de Santa Cruz.
Golfo y costa de la nueva
españa. ca. 1567. Madrid,
Archivo Historico National.**

*The special significance of
this map is that it gives the configuration and names along the
east coast of America from 10°
N.L. to 44° N.L. as they were
known to and accepted by
Spanish cartographers in the
latter half of the sixteenth
century.*

*The pen-and-ink drawing on
paper was found in Seville
among the papers of Santa
Cruz, archicosmographer to
Philip II, following his death on
November 9, 1567. Since "r. de
s. agostino" appears on the
Florida coast and Pedro
Menéndez de Avilés did not
found St. Augustine until the
latter half of 1565, and since
several names along the
Georgia-South Carolina coast
derive from Menéndez's voyage
of 1566, the composition of the
map in Seville could not have
been earlier than 1566 or later
than November, 1567.*

*Santa Cruz gave a new name,
"bª de nrª senora" (Bay of Our
Lady), to St. Mary's Bay,
omitted all islands in the bay,
and gave it a narrow entrance;
he was evidently influenced by
reports of Chesapeake Bay
made by the Spanish in the
early 1560s. North of S. Juan
(Cape Henlopen), however, he
omitted all Gómez names south
of Cape Cod Bay (bª de s.
xpºbal); this was a coast unfre-
quented by Spanish expeditions.*

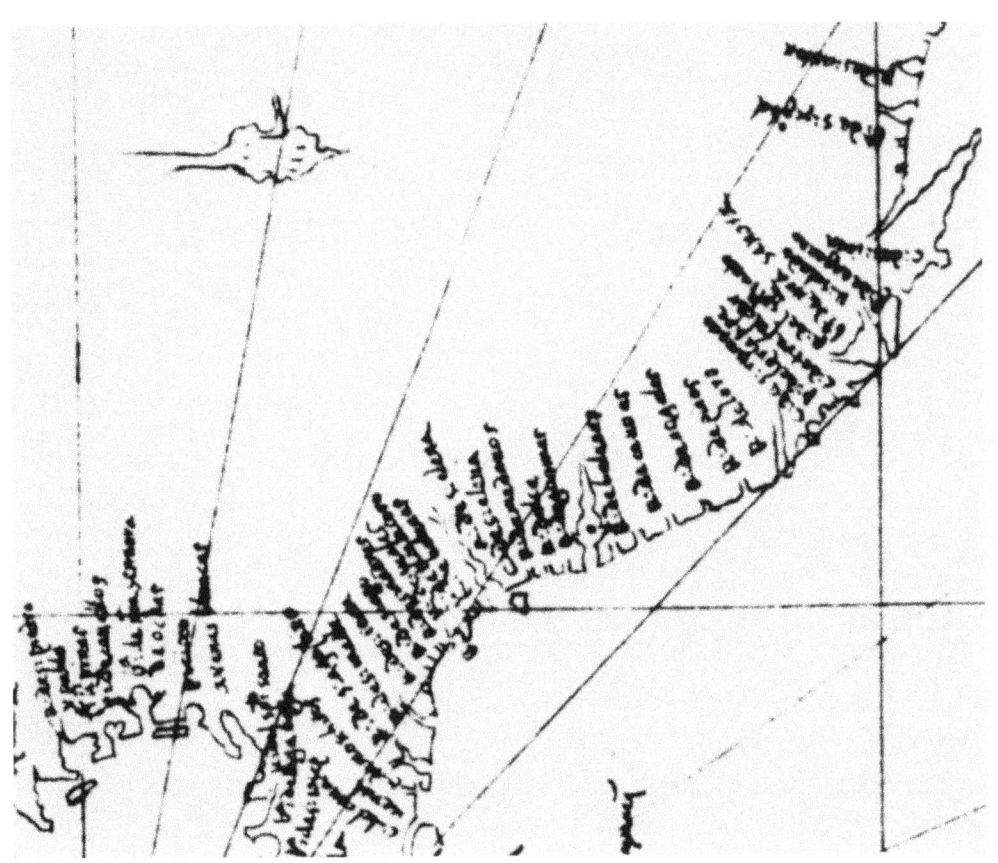

119

PLATE 17.
Gerhardus Mercator.
World Map. 1569. Rotter-
dam. Prins Henrik
Maritiem Museum.

Mercator's engraved world map of 1569 is one of the most important and influential in the history of map-making; this importance is not based on the map's topographical accuracy, however, but on the value of the projection that Mercator evolved for the basically impossible task of portraying the spherical earth accurately on a flat surface. Mercator produced a chart on which a straight line joining any two points on the chart determined the compass direction a navigator must follow in sailing the most direct route from one place to another. Mercator's projection revolutionized navigation; it enabled the mariner to follow a set compass course and reach his destination. The projection also is conformal; landmasses retain their true shape. However, the nearer the poles, the larger is the area scale, because on a globe latitudinal lines converge to a polar point—but on the Mercator projection they remain parallel; meridian lines are as wide at the pole as at the equator. The cylindrical shape of the map results in increasing the size of an area the farther it is from the equator; the best-known illustration is that Greenland on a Mercator map is larger than South America, although it is actually only one-ninth the size.

Mercator's map is a learned compilation from many sources; for the Atlantic coast below the St. Laurence, however, it is dependent chiefly on Spanish sources. At 34° N.L. is a large unnamed islanded bay, evidently Baya de Santa Maria; at 32° is the R. Sola, a great river with its mouth at C. de s. Elena (Port Royal); at C. de arenas (here Cape Cod) is a large, almost landlocked, bay. Mercator's is the first map to show the Appalachians as a continuous mountain range extending parallel to the east coast.

121

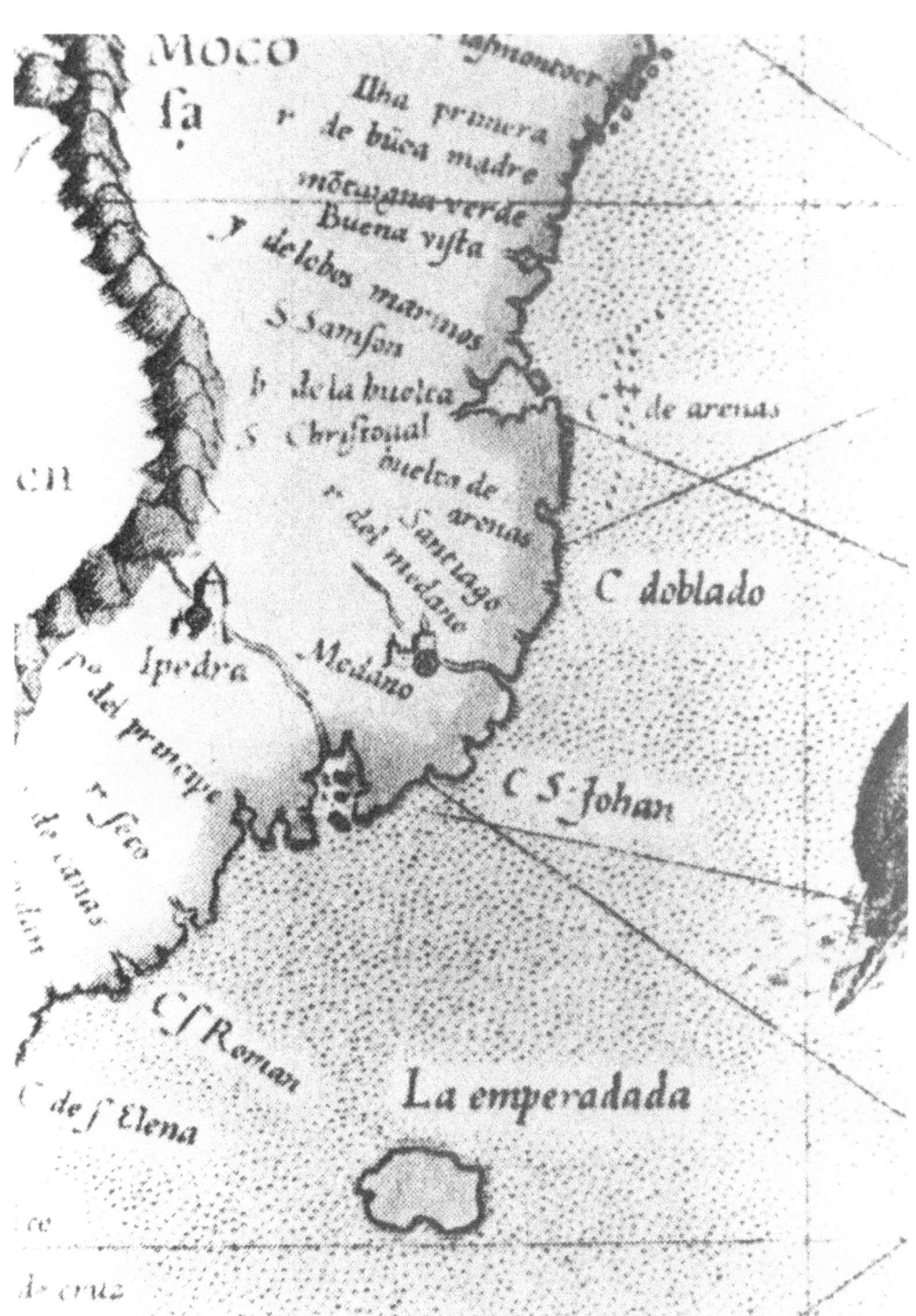

PLATE 18.
John Dee. Atlantidis vulgariter Indiae occidentalis nominatae, emendiator descriptio, quam adhuc est divulgata. 536. London. British Library. Cotton MS, Augustus I. l. I.

This map of eastern North America was prepared from the best sources available to Dr. John Dee, the learned geographer and scientist, and presented by him to the queen at court on October 3, 1580. The coast from the Gulf of Mexico to Cape Breton is based on a careful appraisal of good Spanish charts; the St. Lawrence Gulf and River follow closely Mercator 1569 or a source common to both. The "G[olfo] de S. m^a ," with its narrow, island-blocked entrance, is at 35°30'; Port Ferdinando, at 35°58', was the entrance used by the Roanoke colonists.

Dr. Dee's activities as horoscope caster for the queen's coronation, astrologer, and alchemist resulted in the destruction of his house, possessions, and laboratories by the populace, who thought him a magician who had sold himself to the devil; only recently has his place as learned geographer, respected research chemist, and consultant to powerful Elizabethan leaders been reassessed.

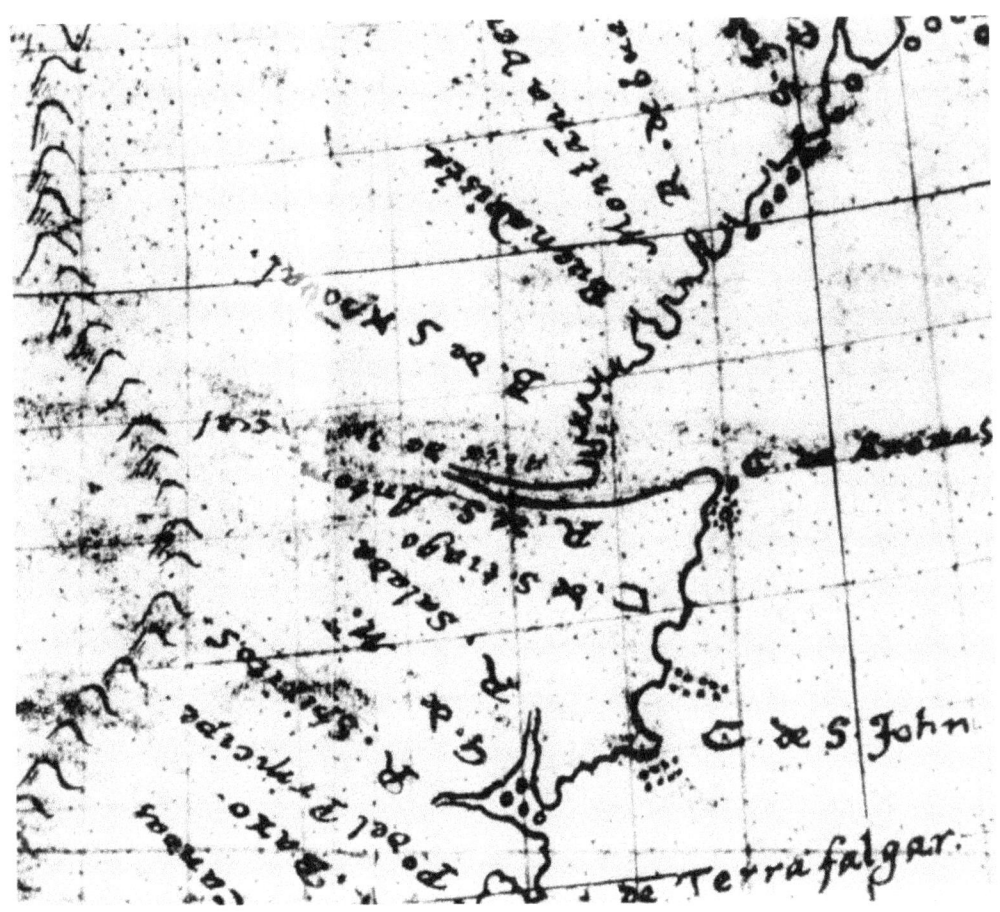

123

PLATE 19.
Simon Fernándes-John Dee. Chart of the Atlantic Coasts. 1580. London. British Library. MS Cott. Rol. XIII. 48.

This vellum chart of the Atlantic and Pacific seacoasts bordering North and South America is a copy, made by Dr. John Dee's assistant, of a large map brought to Dee's residence on November 20, 1580, by Simon Fernándes, a Portuguese pilot who was at the time in the service of Sir Humphrey Gilbert. Fernándes became the pilot major of Ralegh's expeditions to Virginia. His original may be the map from which Governor Ralph Lane, upon arrival in 1585, deduced that Wococon (Ocracoke Inlet) was the "Porte in ye Carte [map] by ye Spanyardes called St Marryes baye." On the copy made by Dr. Dee's assistant, the entrance of Bahia de Santa Maria is between 35°20' and 35°30'; on White's map (plate 23) the inlet to the south of Wococon Island is 35°17'; the actual present latitude of Ocracoke Inlet is 35°04'.

Fernándes's map shows graphically a major reason for the choice of the Carolina Banks for the planting of Ralegh's colony. Santa Maria is the only bay on the coast between the Cape of Florida and Cape Cod Bay and is depicted as a large, open bay with two rows of three islands each across the entrance. It is strategically located for preying on Spanish shipping, and for this profitable Elizabethan sport Fernándes had an obsession, as did most seafaring Englishmen.

125

PLATE 20.

Humphrey Gilbert-John Dee. Polar Chart of the Northern Hemisphere. ca. 1582. Philadelphia. Philadelphia Free Library, William L. Elkins Collection, No. 42.

In this chart of the earth north of the twentieth latitude are portrayed current beliefs and theories of possible water routes above and through North America to the Orient. Verrazzano's Sea's nearest approach to the Atlantic is at 38° N.L., with a wide water passage to the St. Lawrence River. Below Verrazzano's Sea is a large lake with a possible opening to or connection to it; the lake also has an exit to the Atlantic through "R. Meo," which emerges at or near Port Royal, and two wide waterways, one to the Gulf and another to Mexico City and its lake. John White probably saw this map before making his general map of the Southeast.

In the lower right-hand corner is the signature "Humfray Gylbert knight his charte"; below it and elsewhere are three cabalistic signs identifying the author as Dr. John Dee. Both men were deeply interested in water passages for discoveries into and through North America: Sir Humphrey Gilbert as the great navigator and colonial planner about to depart in 1683 on his last tragic voyage, from which he never returned, and Dr. Dee, the philosophical geographer and adviser to colonial expansionists.

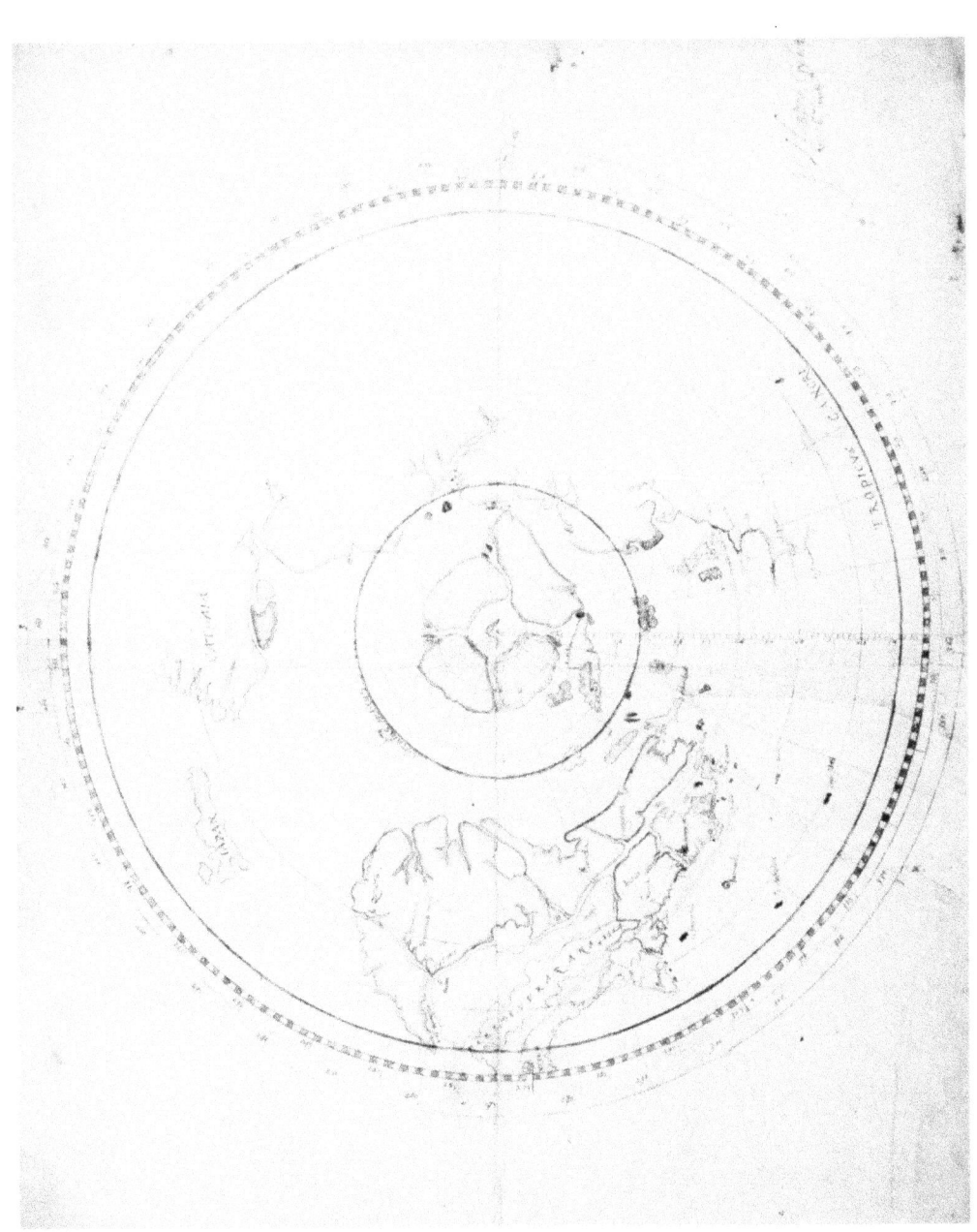

PLATE 21.
Michael Lok: Map of the Northern Hemisphere. 1582.

Lok's map was published by Richard Hakluyt in Divers voyages touching the discoverie of America *(London, 1582) to support English claims to North America and to illustrate routes from the north Atlantic to the Pacific. The Lok map is a catalog of discoveries in North America since the end of the fifteenth century, with names of discoverers and dates of discovery noted. Hakluyt wrote that Lok had an "olde excellent mappe" by Verrazzano; the latitude of the "little necke of lande" at 40°N.L. shows that his Verrazzano map had, like Gerolamo's (plate 5), presently in the Vatican Library, a coastal latitude 5 degrees too high. The Verrazzanian geography on this map and on that of the Gilbert-Dee chart gave to the Englishmen of the period, especially the colonial planners, a strong mental picture of North American colonization that included, in Hakluyt's words (in his "Particuler discourse" of 1584), the belief that "the North-west passage to Cathaio and China may . . . be searched out as well by River and overlande as by sea."*

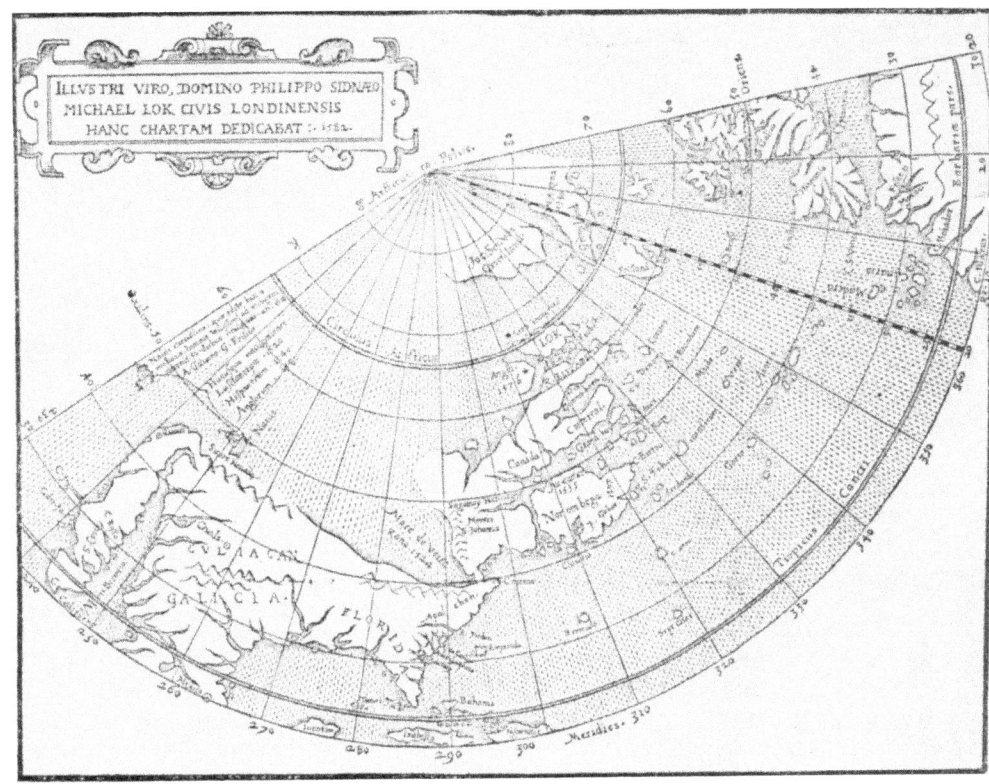

129

PLATE 22.

Anonymous. A discription of the land of virginia [endorsed]. London. Public Record Office, Maps and Plans G. 584 (formerly CO. 1/5, 42).

This untitled, unsigned, undated rough sketch in the Public Record Office was identified by historian David Beers Quinn as a drawing of early explorations made by the Ralegh colonists in the summer of 1585. It was probably enclosed by Governor Lane in his letter of September 8, 1585, to Sir Francis Walsingham, principal secretary to the queen and a patron of the colony, and sent on the return voyage of the Roebuck *or* Elizabeth. *Its accuracy of topographical relationships indicates that it may be a copy from early surveys by the skilled mathematician Thomas Harriot. The legends on the map are transcribed in clockwise order from the bottom.*

1. *the port of saynt maris wher we arivid first: [This is the Bay de Santa Maria, identified on the Spanish map to which Governor Lane refers.]*
2. *Wococon: [Ocracoke Island]*
3. *here groith ye roots that diethe read: [dogwood bark?]*
8. *pomaioke: [visited by colonists July 12; behind it is Lake Paquippe]*
6. *here is .3. fatham of water: [Grenville and party reached Pamlico River on July 15]*
7. *secotan: [an important Indian village]*
5. *[t]his goithe to a great toune called nesioke: [Indian town on Core Sound, visited July 16?]*
4. *[   ] to warreā: [probably visited on return trip from Pamlico River, July 17, 1585]*
9. *freshe water with great store of fishe: [Albemarle Sound with the Roanoke and Chowan rivers, explored by Amadas September 2-7]*
10. *here were great store of great red grapis very pleasant: [see grapes (Muscadines?) in de Bry's pictorial engraving]*
11. *the grase that berither the silke groithe here plentifully: [fibre-yielding plants on Croatan Island]*
12. *y^e kinges ill: [Roanoke Island; the "king" was Wingina, later the colonists' bitter enemy]*
13. *the gallis are found here: [oak-galls, for tanning; or Rosegentian, a medicinal plant?]*

131

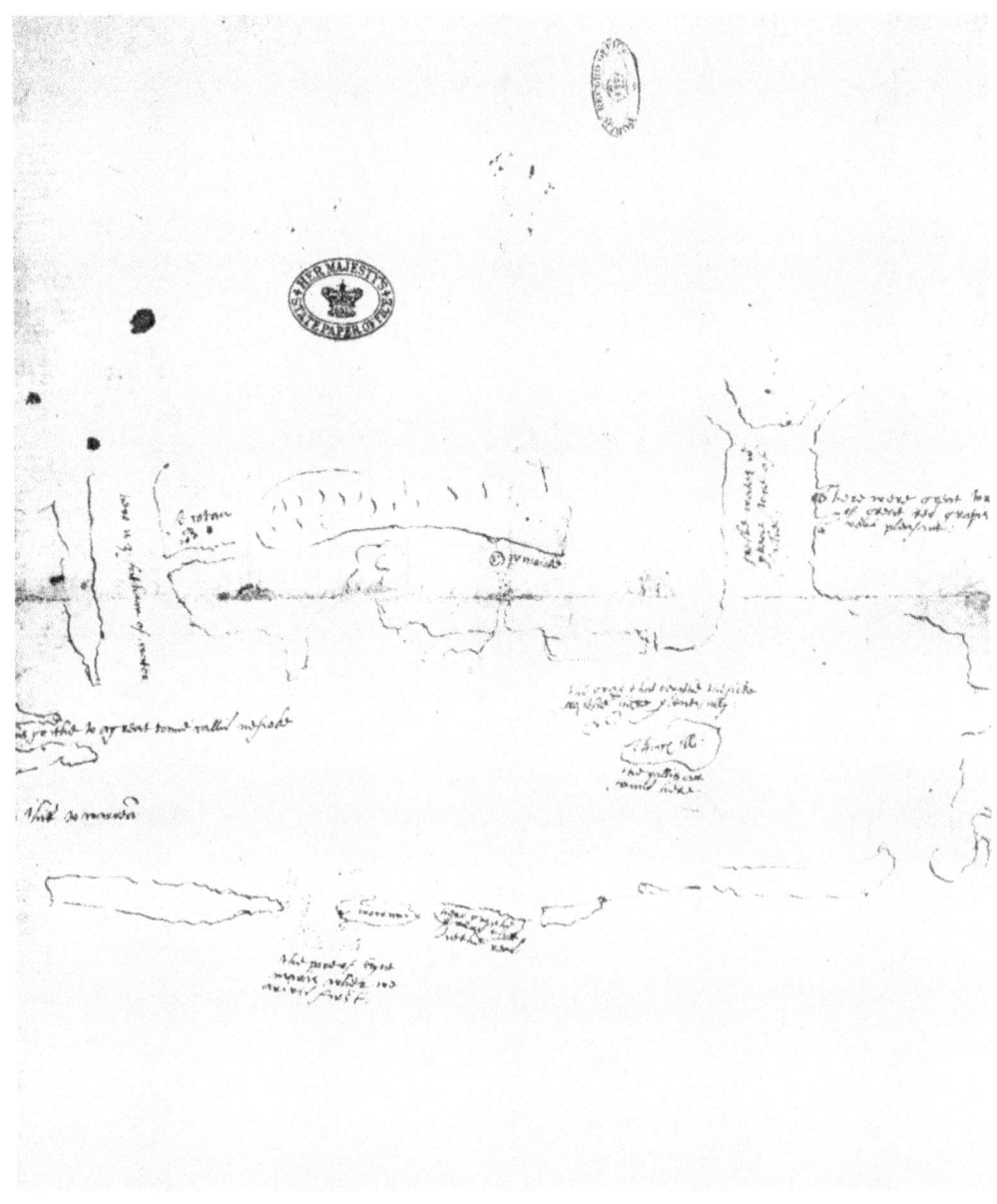

PLATE 23.
John White. La Virgenia Pars. 1585. London. British Library, Prints and Drawings, 1906-5-9-1(2), L.B.1.

John White, an artist who accompanied the first English colonists to Roanoke, later became governor of the settlement. A portfolio of his drawings is preserved in the British Library. Among them is this general map of the Florida peninsula and the east coast from Florida to the southern shore of an unnamed bay, the Chesapeake. In addition to the North Carolina Outer Banks, an area explored and surveyed by the English, the map shows the influence of earlier Spanish and French conceptions of the coast south of the Outer Banks.

Some of the sources available to and used by White may be identified. Fernándes's own copy of the Spanish-type map copied by Dee's assistant (plate 19) is the apparent basis for the Bahama Islands; le Moyne's map of Florida and the east coast to Port Royal, in an earlier version than the one used for de Bry's engraving in 1591, is the obvious source for that area; the strait or river from Port Royal to a large body of water (Verrazzano's sea?) closely follows the so-called Harleian map (plate 12); the Florida west coast is based upon an unidentified Spanish map. Between Port Royal and Cape Lookout are several landmarks such as present Cape Fear, Winyah, and Charleston bays, which the English probably noted as they sailed north from the West Indies, as they are more clearly identifiable here than in the coastal topography of any extant Spanish map. White's map is valuable as a compilation of what the Roanoke colonists knew or thought they knew.

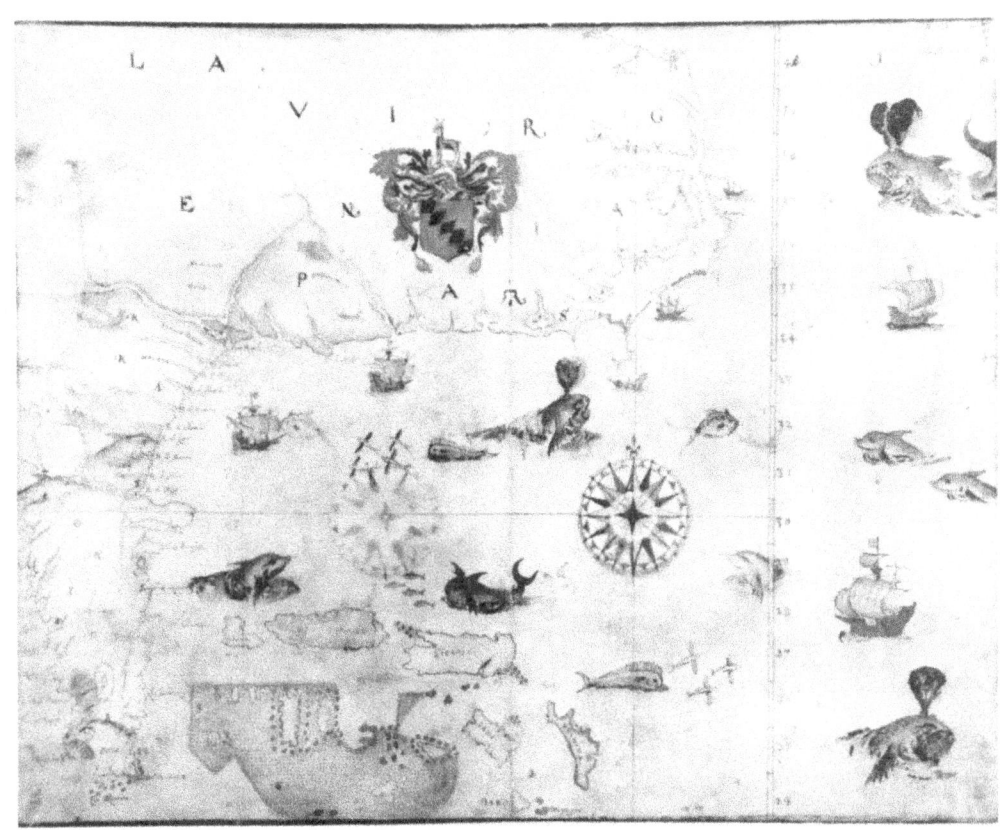

133

PLATE 24.
John White. Virginea Pars. 1585. London. British Library, Prints and Drawings, 1906-5-9-1(3), L.B.1(a).

White's careful drawing of the region between Cape Lookout and Chesapeake Bay, both unnamed on the map, is the most authoritative preserved document showing the location and names of Indian villages and the geography of the Outer Banks and area surrounding the sounds as they had been surveyed by or known to the Roanoke colony. White's fellow colonist, the brilliant mathematician and teacher of navigation, Thomas Harriot, is almost certainly responsible for the surprising accuracy shown in the representation of so large an area, one of several hundred square miles, made with considerable limitations of time in gathering information as well as of personnel and surveying equipment available.

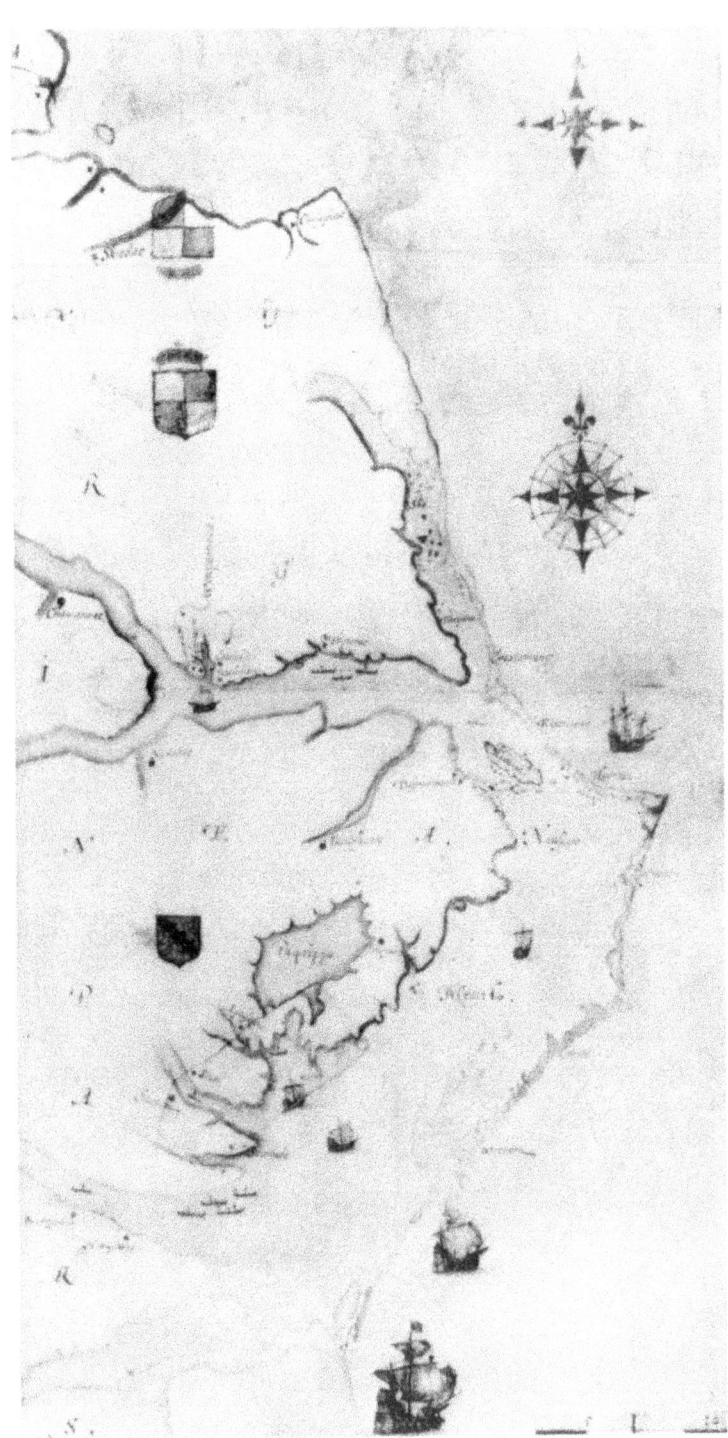

PLATE 25.
John White-Theodor de Bry. Americâe pars, Nunc Virginia dicta. 1590. Engraved in Theodor de Bry, *America,* part 1, No. 1. Frankfurt-am-Main, 1590.

De Bry's engraving of White's drawing of Ralegh's Virginia extends from an unnamed inlet below "Promontorium tremendum" (present Cape Lookout) to Chesapeake Bay. Minor differences in the topography, and in the names of Indian villages added to as well as eliminated from those on the White manuscript map, indicate that de Bry worked from a different and probably later revised version. De Bry's engraving provided European geographers and mapmakers with a delineation of the coast that was more accurate and detailed than any previous work; it became a prototype not only for maps of the Southeast but also, by the middle of the seventeenth century, for western hemispheric and world maps.

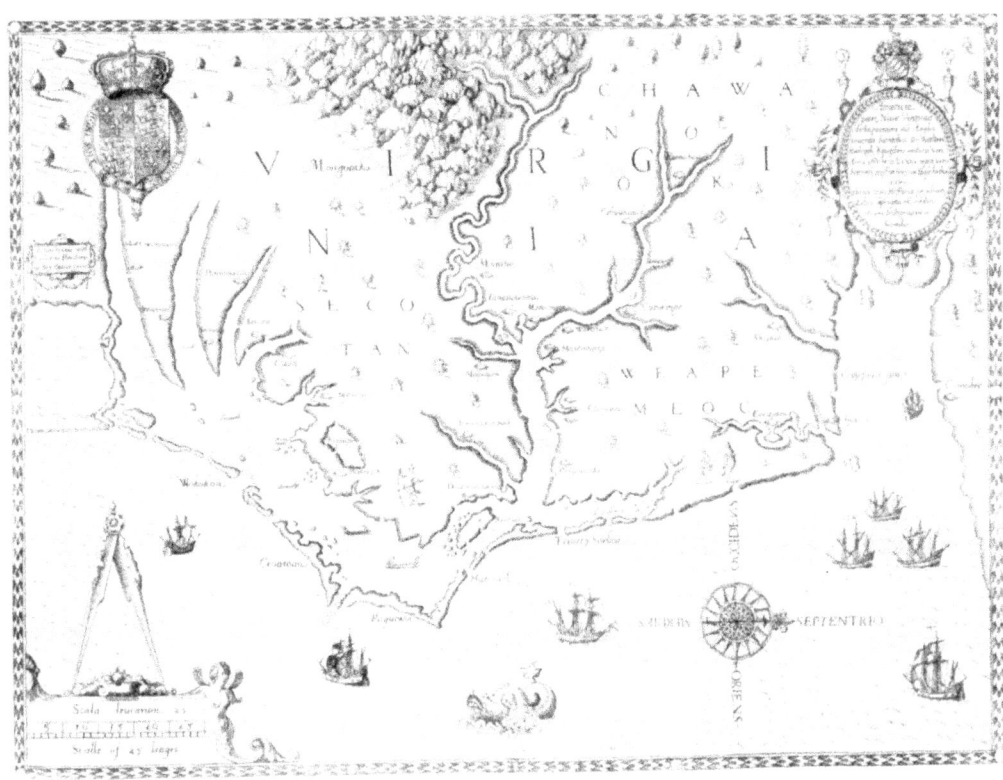

PLATE 26.

PLATE 26.
John White-Theodor de Bry. The arriual of the Englishemen in Virginia. Engraved in Theodor de Bry, *America*, Part 1. Frankfurt-am-Main, 1590, plate 2.

This semipictorial map portrays with considerable elaboration the northern half of the area shown in the rough sketch map sent to England soon after the arrival of the first colony in 1585. White's original drawing is lost. De Bry engraved this somewhat conventionalized offshore view of the country with palisaded Indian villages; with over half a dozen varieties of trees, including grapevines and live oaks with Spanish moss; with Indians poling their dugouts; and with a sunken ship at each inlet to show the dangers of their shallowness. None of these details are on the manuscript maps.

Below the engraving is an account of the 1584 landing that is somewhat different from Barlowe's printed version of the arrival of the English. Written by Thomas Harriot in Latin, it was translated for this English edition by Richard Hakluyt the younger.

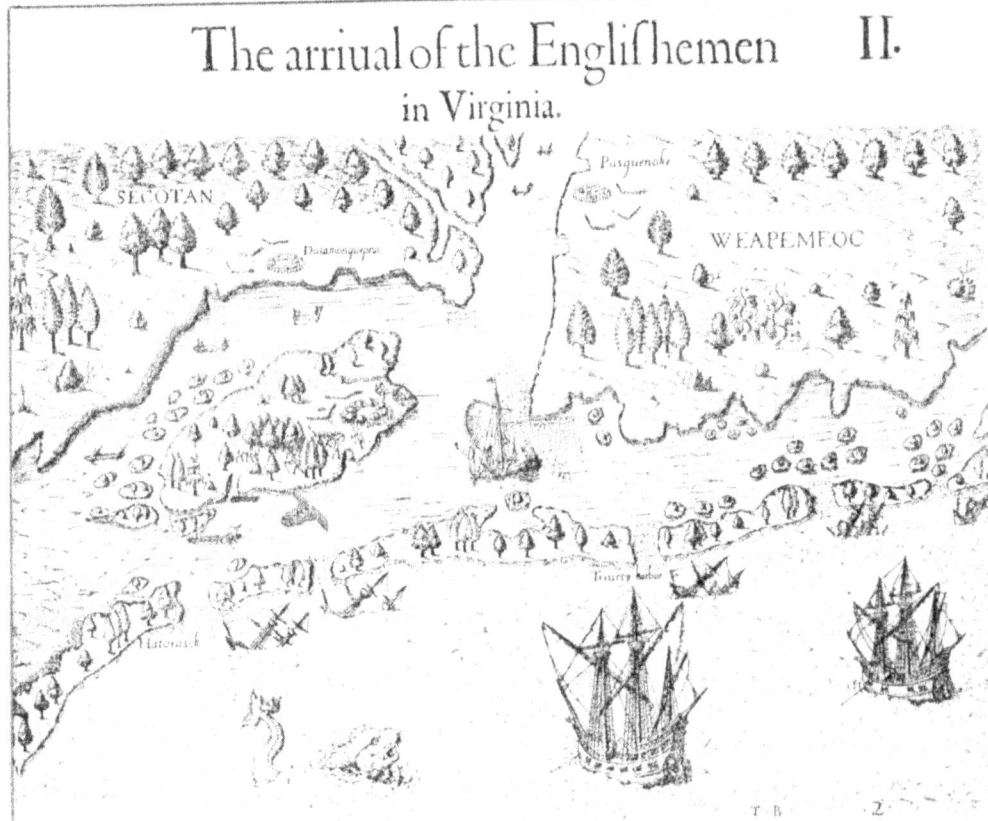

PLATE 27.

Gerhardus Mercator-Jodocus Hondius. Virginiae Item et Floridae Americae Provinciorum nova Descriptio. 1606. From Gerhardus Mercator. *Atlas . . . auctus ac.* **illustratus á Judoco Hondio. Amstelodam, 1606, no. 143.**

This is one of the most influential and beautifully designed maps of the Southeast; it was designed and engraved by Jodocus Hondius, who acquired the map plates of Mercator and issued his atlas with fifty new plates by himself and others. Hondius based his map on a skillful combination of two maps published by de Bry: those of White's Virginia, 1590, and of le Moyne's Florida, 1591. Subsequent cartographers usually followed this work rather than the two earlier maps; it thus became the most important type map of the region until the Carolina Lords Proprietors maps by Ogilby (ca. 1672) and Gascoyne (1682). Even in the eighteenth century, atlases of the great European publishers continued to use plates of the region between Virginia and the Florida peninsula that followed closely the Mercator-Hondius map of 1606.

PLATE 28.

Pedro de Zuñiga. [Chart of Virginia]. 1608. Simancas, Spain. Archivo General de Simancas. M.P.D., IV-66, XIX-163.

This chart of the region surrounding Chesapeake Bay has three rivers to the south of James River that represent the Neuse, Pamlico, and Roanoke. The map is so crudely drawn that the Outer Banks are omitted. The map has numerous legends, almost illegible; in notes on the map are references to survivors of the Roanoke colony and to the continued belief that the South Sea (the Pacific) is so near to the headwaters of the James River that the salt waves dash into it.

The chart is drawn in an early seventeenth-century English hand by Captain John Smith or copied from his map; it was obtained by Pedro de Zuñiga, ambassador to England, and sent to Spain in September, 1608.

143